ROUTLEDGE LIBRARY EDITIONS:
20TH CENTURY SCIENCE

T0250306

Volume 18

THE GENERAL PRINCIPLES OF QUANTUM THEORY

THE GENERAL PRINCIPLES OF QUANTUM THEORY

G. TEMPLE

Routledge
Taylor & Francis Group
LONDON AND NEW YORK

First published in 1934
Second edition published in 1942
Third edition published in 1945

This edition published in 2014
by Routledge
2 Park Square, Milton Park, Abingdon, Oxfordshire OX14 4RN

and by Routledge
711 Third Avenue, New York, NY 10017

Routledge is an imprint of the Taylor andFrancis Group, an informa business

First issued in paperback 2015

British Library Cataloguing in Publication Data
A catalogue record for this book is available from the British Library

ISBN 978-0-415-73519-3 (Set)
eISBN 978-1-315-77941-6 (Set)
ISBN 978-1-138-01364-3 (hbk) (Volume 18)
ISBN 978-1-138-99786-8 (pbk) (Volume 18)
ISBN 978-1-315-77922-5 (ebk) (Volume 18)

Publisher's Note
The publisher has gone to great lengths to ensure the quality of this book but points out that some imperfections from the original may be apparent.

Disclaimer
The publisher has made every effort to trace copyright holders and would welcome correspondence from those they have been unable to trace.

The General Principles of Quantum Theory

G. TEMPLE
Ph.D., D.Sc.

LONDON: METHUEN & CO LTD
NEW YORK: JOHN WILEY & SONS INC

First published 1934
Reprinted seven times
Reprinted 1961

5.4

CATALOGUE NO. (METHUEN) 2/4012/11

REPRINTED BY LITHOGRAPHY IN GREAT BRITAIN
BY JARROLD AND SONS LIMITED, NORWICH

PREFACE

THE object of this monograph is to give an introductory account of the general principles which form the physical basis of the Quantum Theory. This theory is here considered as a branch of physics and not as a branch of mathematics; hence questions of mathematical technique are treated only in relation to the appropriate expression of physical concepts in mathematical language. The exposition is restricted to a discussion of general principles and does not attempt their detailed application to the wide domain of atomic physics, although a number of special problems are considered in elucidation of the general principles. Numerous examples are given to illustrate the general theory and to indicate the nature of further developments outside the scope of this book.

The necessary fundamental mathematical methods—the theory of linear operators and of matrices—are developed *ab initio* in the first chapter. The method of Wave Mechanics, which forms the subject of Dr. H. T. Flint's monograph in this series, is here considered only as that form of the general mathematical method appropriate to variables with continuous spectra.

In a short introductory work such as this it appears unnecessary to burden the text with detailed references to original authorities. My debt to the pioneers of the subject is indicated by the general references at the end of this book, and I especially wish to acknowledge my indebtedness to the standard treatises of Dirac, Weyl and J. v Neumann.

G. T.

HINDHEAD, *August, 1933*

PREFACE TO SECOND EDITION

No extensive changes, but a number of minor corrections, have been made in the second edition.

G. T.

April, 1942

PREFACE TO THIRD EDITION

I AM grateful to all the readers who have enabled me to correct a number of misprints and small errors.

G. T.

6 February, 1945

CONTENTS

THE GENERAL PRINCIPLES OF QUANTUM THEORY

CHAPTER I

THE THEORY OF LINEAR OPERATORS

THE principal mathematical instrument in the formation and development of the quantum theory is the theory of linear operators. This branch of pure mathematics plays the same part in the quantum theory as tensor analysis in the theory of relativity, or the infinitesimal calculus in classical dynamics. A study of this subject is therefore an indispensable preliminary to the study of the quantum theory, and although the theory of the present chapter may be unfamiliar to many students of physics, its concepts and methods are essentially simple and find wide applications in both mathematics and physics.

Complex Numbers as Operators.—A simple introduction to the theory of linear operators is provided by the geometrical interpretation of ordinary complex numbers as transformations of a coplanar set of vectors. A vector is completely specified by its components parallel to the axes, which we suppose to be rectangular. It is convenient to change the usual notation in anticipation of the subsequent generalisation of the theory to n dimensions, and to denote the components of the vectors α and β by a_1, a_2 and b_1, b_2, etc. The following definitions complete the theory of plane vectors. The sum of two vectors α and β is the vector $\alpha + \beta$ with components $a_1 + b_1$, $a_2 + b_2$. The product of a vector α by a number c is the vector $c\alpha = \alpha c$ with components ca_1, ca_2. The unit vectors along the axes will be denoted by ϵ_1 and ϵ_2, so that $\alpha = \epsilon_1 a_1 + \epsilon_2 a_2$. The magnitude of the vector α is $(a_1^2 + a_2^2)^{\frac{1}{2}}$, and the scalar product of α and β is $(a_1 b_1 + a_2 b_2)$, written for brevity as (α, β).

Complex numbers are entities essentially different from real numbers and from those ordered pairs of real numbers which we call vectors. These latter represent magnitudes and positions respectively—i.e. they are passive or static in character: but complex numbers represent operations executed upon vectors—i.e. they are active or dynamic. Briefly, real numbers and vectors are nouns, complex numbers are verbs—in the imperative mood! Unfortunately this vital distinction is obscured in current mathematical symbolism which uses the same symbols, 1, 2, . . . x, for ordinary real numbers and for complex numbers whose imaginary part is zero. To emphasise the essential difference between these two concepts we shall use I for the complex number usually denoted by $1 + 0i$, and J for $0 + 1i$.

The symbols I and J represent the two fundamental transformations in the theory of complex numbers. I represents the "identical transformation" which leaves every vector unaltered, and J represents a counter-clockwise rotation through one right angle. This transforms a vector α, with components a_1, a_2, into a vector $\alpha' = \mathrm{J}\alpha$, with components $-a_2$, a_1. To these symbols we may add O representing the "nul transformation" which annihilates every vector, i.e. replaces it by the nul-vector with zero components. Of course we must distinguish between the passive bourgeois zero 0 and this active, nihilist zero O. The symbols I, J, O and all similar symbols representing transformations of vectors are called "operators."

The transformation symbolised by J can be repeated. It then transforms the vector α', with components

$-a_2$, a_1, into the vector α'', with components $-a_1$, $-a_2$, i.e. the vector $-\alpha$. The repetition of an operation is represented symbolically by writing

$$\alpha'' = J\alpha' = J \,.\, J\alpha = J^2\alpha,$$

whence $J^2\alpha = -\alpha$. This justifies the definition of J as the imaginary unit in the algebra of complex numbers.

To define the general complex number, with real part x and imaginary part y, we consider the transformation which replaces the vector α, with components a_1, a_2, by the vector α', with components

$$a_1' = xa_1 - ya_2,$$
$$a_2' = ya_1 + xa_2.$$

If Z is the operator representing this transformation, then it is clear that

$$\alpha' = Z\alpha = xI\alpha + yJ\alpha,$$

and Z is therefore defined to be the complex number usually written as $x + yi$. The complete justification of this identification requires definitions of the sum and product of two operators, which are given in the next section.

Two-Dimensional Operators in General.—It is clear that besides operators like Z, which represent complex numbers, there are many other types of operators acting on the vectors in a given plane. Of these the simplest are the "projective operators" defined as follows: If π is any unit vector, the projection of α on π has the magnitude (π, α) and the direction of π, and is therefore equal to $\pi(\pi, \alpha)$. The operator P defined by the equation

$$P\alpha = \pi(\pi, \alpha),$$

is called the projective operator for the unit vector π. These operators possess the important property that $P^2\alpha = P\alpha$. For, since $P\pi = \pi$,

$$P \,.\, P\alpha = P\pi(\pi, \alpha) = \pi(\pi, \alpha) = P\alpha.$$

Another simple but much less important type of operator is that which "reflects" a vector in a prescribed line. E.g. the equations

$$a_1' = a_1, \quad a_2' = -a_2,$$

and

$$a_1' = a_2, \quad a_2' = a_1,$$

respectively specify operators L and M which reflect the vector α in the x-axis and in the line whose equation is $x = y$.

In general an operator R is completely specified by the equations expressing the components of $\alpha' = \mathrm{R}\alpha$ in terms of the components of α. If these equations are linear, R is called a "linear" operator and then the equations have the form

$$a_1' = r_{11}a_1 + r_{12}a_2,$$
$$a_2' = r_{21}a_1 + r_{22}a_2.$$

The coefficients r_{jk} are independent of a_1, a_2 and the complete set of these coefficients $\begin{Vmatrix} r_{11} & r_{12} \\ r_{21} & r_{22} \end{Vmatrix}$ is called the "matrix" of R. Thus, for the projective operator P, the "matrix element" $p_{jk} = p_j p_k$, where p_1, p_2 are the components of the unit vector π associated with P. Similarly, the matrices of the operators I, J, L and M are

$$\begin{Vmatrix} 1 & 0 \\ 0 & 1 \end{Vmatrix}, \quad \begin{Vmatrix} 0 & -1 \\ 1 & 0 \end{Vmatrix}, \quad \begin{Vmatrix} 1 & 0 \\ 0 & -1 \end{Vmatrix}, \text{ and } \begin{Vmatrix} 0 & 1 \\ 1 & 0 \end{Vmatrix}.$$

The sum of two operators R and S is defined to be the operator R + S such that, for any vector α,

$$(\mathrm{R} + \mathrm{S})\alpha = \mathrm{R}\alpha + \mathrm{S}\alpha.$$

The product of an operator R and a real number c is defined to be the operator cR such that, for all vectors α,

$$(c\mathrm{R})\alpha = c(\mathrm{R}\alpha).$$

Thus we may write $\mathrm{Z} = x\mathrm{I} + y\mathrm{J}$, suppressing the vector operand α. Similarly, the product of an operator

R by an operator S is defined to be the operator SR such that, for any vector α,

$$(\mathrm{SR})\alpha = \mathrm{S}(\mathrm{R}\alpha),$$

i.e. $(\mathrm{SR})\alpha = \gamma$, where $\gamma = \mathrm{S}\beta$ and $\beta = \mathrm{R}\alpha$. Thus we find that $\mathrm{JL} = \mathrm{M}$ and $\mathrm{LJ} = -\mathrm{M}$, a result which shows that the product RS is not necessarily the same as the product SR. The multiplication of operators is therefore not commutative in general, and the *order* of factors in a product is vitally important.

The relations connecting the matrix elements of R and S with the matrix elements of the sum $\mathrm{U} = \mathrm{R} + \mathrm{S}$ and the product $\mathrm{V} = \mathrm{SR}$ follow at once from the preceding definitions. They are

$$u_{jk} = r_{jk} + s_{jk},$$

and

$$v_{jk} = s_{j1}r_{1k} + s_{j2}r_{2k}.$$

A simple illustration of these definitions is provided by the addition and multiplication of the operators, $\mathrm{Z} = x\mathrm{I} + y\mathrm{J}$ and $\mathrm{W} = u\mathrm{I} + v\mathrm{J}$. It is found that

$$\mathrm{W} + \mathrm{Z} = (u + x)\mathrm{I} + (v + y)\mathrm{J},$$

and

$$\mathrm{WZ} = (ux - vy)\mathrm{I} + (uy + vx)\mathrm{J} = \mathrm{ZW}.$$

These results are the familiar laws for the addition and multiplication of complex numbers, and they complete the justification of the definition given in the preceding section.

Proper Vectors and Proper Values in Two Dimensions.—In general the action of an operator R on a vector α transforms it into a vector α' which differs from α in magnitude and direction. Nevertheless, it is sometimes possible to find certain vectors which are changed only in magnitude but not in direction by the action of R. These vectors are called the "proper vectors" of R, and the ratios in which their magnitudes are changed by R are called the "proper values" of R.

For example, the reflection operator L has the proper vectors ϵ_1 and ϵ_2 with proper values 1 and -1; the operator M has the proper vectors $\epsilon_1 + \epsilon_2$ and $\epsilon_1 - \epsilon_2$

with proper values 1 and -1; the projection operator P has as proper vectors π and a vector ν perpendicular to π, with proper values 1 and 0.

Turning to the general operator R defined above (p. 4), it is clear that α will be a proper vector of R with proper value r if

$$ra_1 = a_1' = r_{11}a_1 + r_{12}a_2,$$

and

$$ra_2 = a_2' = r_{21}a_1 + r_{22}a_2.$$

Hence r must satisfy the equation

$$(r_{11} - r)(r_{22} - r) = r_{12}r_{21},$$

i.e.

$$2r = r_{11} + r_{22} \pm [(r_{11} - r_{22})^2 + 4r_{12}r_{21}]^{\frac{1}{2}}.$$

If these values of r are real and distinct, say r_1 and r_2, then R will have two proper vectors, ρ_1 and ρ_2.

In the case of the operators J and Z the equations for r are $r^2 + 1 = 0$ and $(r - x)^2 + y^2 = 0$, so that J and Z have no real proper vectors (unless $y = 0$). To avoid the inconvenience of this result we must generalise the theory so as to admit vectors with complex components, i.e. we must deal with vectors whose components are themselves operators. This generalisation is carried out in the next section where also the number of dimensions is increased from 2 to n.

EXAMPLES :—

(1) Show that the matrix $\left\|\begin{matrix} \frac{1}{2} & \frac{1}{2} \\ \frac{1}{2} & \frac{1}{2} \end{matrix}\right\|$ represents a projective operator, and find its proper vectors. [$\epsilon_1 \pm \epsilon_2$.]

(2) If the scalar product (ρ_1, ρ_2) of the proper vectors of R is zero, and the proper values of R are distinct, prove that $r_{12} = r_{21}$, i.e. the matrix of R is symmetric about the leading diagonal.

(3) If P_1 and P_2 are the projective operators for the proper vectors ρ_1 and ρ_2 of R and if $(\rho_1, \rho_2) = 0$, prove that

$$P_1 + P_2 = I,$$
$$P_1P_2 = O = P_2P_1,$$
$$r_1P_1 + r_2P_2 = R,$$
$$r_1{}^2P_1 + r_2{}^2P_2 = R^2.$$

(4) Show that R satisfies the equations

$$\begin{vmatrix} r_{11} - \mathrm{R} & r_{12} \\ r_{21} & r_{22} - \mathrm{R} \end{vmatrix} = (\mathrm{R} - r_1)(\mathrm{R} - r_2) = 0.$$

Vectors in *n* Dimensions.—The generalisation of the preceding theory to n-dimensional space with complex co-ordinates is purely formal. A vector α is now specified by its n components, $a_1, a_2, \ldots, a_n$, which will in general be complex numbers. The sum of two vectors α and β and the product of a vector α by a complex number c are defined as before as the vectors with components $a_1 + b_1, a_2 + b_2, \ldots, a_n + b_n$ and $ca_1, ca_2, \ldots, ca_n$. The principal unit vectors are now denoted by $\epsilon_1, \epsilon_2, \ldots, \epsilon_n$, all the components of ϵ_k being zero except the k-th which is unity. Hence

$$\alpha = \epsilon_1 a_1 + \epsilon_2 a_2 + \ldots + \epsilon_n a_n.$$

To ensure that the magnitude of a vector shall be real and positive a slight change is made in its definition. The magnitude of α is defined as

$$(a_1{}^* a_1 + a_2{}^* a_2 + \ldots + a_n{}^* a_n)^{\frac{1}{2}},$$

$a_k{}^*$ being the complex conjugate of a_k. Similarly, the scalar product of α by β is defined as

$$(\beta, \alpha) = b_1{}^* a_1 + b_2{}^* a_2 + \ldots + b_n{}^* a_n.$$

We note that (β, α) and (α, β) are conjugate complex numbers, i.e. the order of the factors in a scalar product is now important.

The theory of vectors is completed by the following definitions and theorems: α is said to be a unit vector if $(\alpha, \alpha) = 1$. Two vectors, α and β, are said to be orthogonal if $(\alpha, \beta) = 0$. (This equation obviously implies that $(\beta, \alpha) = 0$.) Any set of n unit, orthogonal vectors, $\alpha_1, \alpha_2, \ldots, \alpha_n$, is called a "basis," as, for example, the n principal unit vectors, $\epsilon_1, \epsilon_2, \ldots, \epsilon_n$. The principal properties of a basis are as follows:—

I. The n vectors of a basis are linearly independent,

i.e. they are not connected by any effective relation of the form

$$\alpha_1 c_1 + \alpha_2 c_2 + \ldots + \alpha_n c_n = 0.$$

(For since $(\alpha_j, \alpha_k) = 0$ if $j \neq k$, or 1 if $j = k$, such a relation implies that

$$0 = (\alpha_j, \alpha_1 c_1 + \alpha_2 c_2 + \ldots + \alpha_n c_n) = c_j,$$

for all values of j, i.e. all the c_j's are zero.)

II. *The Expansion Theorem:* any vector ϕ can be expressed in the form

$$\phi = \alpha_1 c_1 + \alpha_2 c_2 + \ldots + \alpha_n c_n.$$

(For this equation yields n simultaneous equations to determine the c_j's, and by I, the determinant of the coefficients of the c_j's does not vanish. Hence there is a unique solution.)

The values of the c_j's are easily found explicitly, for

$$(\alpha_j, \phi) = \sum_k (\alpha_j, \alpha_k) c_k = c_j.$$

Hence $$\phi = \sum_j \alpha_j (\alpha_j, \phi).$$

III. *The Generalised Theorem of Pythagoras:* if ϕ and ψ are any two vectors, then

$$(\psi, \phi) = \sum_j (\psi, \alpha_j)(\alpha_j, \phi).$$

This follows at once from the preceding theorem.

Operators in *n* Dimensions.—A linear operator R is completely specified by the n equations giving the components of $\alpha' = \mathrm{R}\alpha$ in terms of the components of α. If these (scalar) equations are

$$a_j' = \sum_k r_{jk} a_k, \quad (j = 1, 2, \ldots, n),$$

the array of the n coefficients $\| r_{jk} \|$ is called the matrix of R, and the coefficients are called the matrix elements of R. The operator R can also be specified by the n (vector) equations

$$\mathrm{R}\epsilon_j = \sum_k \epsilon_k r_{kj}, \quad (j = 1, 2, \ldots, n),$$

which determine the action of R on the n principal unit vectors. In writing these equations it is convenient to arrange the factors in each product so as to bring together the "dummy" suffixes with respect to which the summation is effected.

The sum and product of two operators are defined as before. If $U = R + S$ and $V = SR$, the matrix elements of U and V will be found to be

$$u_{jk} = r_{jk} + s_{jk},$$

and

$$v_{jk} = \sum_l s_{jl} r_{lk}.$$

The second equation expresses "the matrix law of multiplication." The matrix elements of R can be expressed as scalar products in the form

$$r_{jk} = (\epsilon_j, R\epsilon_k).$$

These definitions can be illustrated by application to an especially important type of operator, described as symmetric, self-adjoint, or Hermitian by different writers, and most simply defined by the condition that the scalar product $(\alpha, R\alpha)$ is always real for a symmetric operator R. It follows at once that the diagonal matrix elements of R, i.e. those of the form r_{kk}, are all real, and it is easily proved that the matrix of R possesses the limited degree of symmetry expressed by the equation

$$r_{jk} = r^*_{kj}.$$

Further, it can be shown that, if R is symmetric, then for any pair of vectors β and γ,

$$(\beta, R\gamma) = (R\beta, \gamma).$$

Both theorems depend upon the lemma that if $pz + qz^*$ is real for every pair of conjugate complex numbers z, z^*, then p and q must also be conjugate complex numbers. To prove the theorems we choose α to be $\beta z + \gamma z^*$. Then

$$(\alpha, R\alpha) = \text{a real number} + (\beta, R\gamma)z^{*2} + (\gamma, R\beta)z^2.$$

Hence by the lemma

$$(\beta, R\gamma) = (\gamma, R\beta)^*.$$

But $\qquad (R\beta, \gamma) = (\gamma, R\beta)^*,$

therefore $\qquad (\beta, R\gamma) = (R\beta, \gamma).$

Also, on writing $\beta = \epsilon_j$, $\gamma = \epsilon_k$, we find that

$$r_{jk} = (\epsilon_j, R\epsilon_k) = (\epsilon_k, R\epsilon_j)^* = r_{kj}^*.$$

Projective Operators.—The projective operator A which "projects" any vector ϕ on to the unit vector α is defined, as in two dimensions, by the equation

$$A\phi = \alpha(\alpha, \phi),$$

whence it follows, as before, that $A^2 = A$, that A is symmetric, and that the matrix elements of A are of the form

$$a_{jk} = a_j a_k^*.$$

The mathematical expression of the physical principles of the next chapter requires the consideration of the relation between two projective operators A and B with unit vectors α and β. Since

$$B\phi = \beta\,(\beta, \phi) \quad \text{and} \quad A\beta = \alpha(\alpha, \beta),$$

it follows that

$$AB\phi = \alpha(\alpha, \beta)(\beta, \phi),$$

and $\qquad BAB\phi = \beta(\beta, \alpha)(\alpha, \beta)(\beta, \phi) = (\beta, \alpha)(\alpha, \beta)B\phi,$

i.e. $\qquad BAB \equiv cB, \quad \text{where} \quad c = |(\alpha, \beta)|^2.$

Similarly, $\qquad ABA \equiv cA.$

Hence $c^{-1}AB$ and $c^{-1}BA$ are also projective operators for the unit vectors α and β.

If α and β are orthogonal, then

$$AB = O = BA,$$

and the projective operators A and B are also said to be orthogonal. If every pair of the n vectors, α_1, α_2,

. . ., α_n, are orthogonal, the same is true of the n associated projective operators, A_1, A_2, . . ., A_n, i.e.

$$A_jA_k = O \quad \text{if} \quad j \neq k, \qquad (1)$$
$$= A_k \quad \text{if} \quad j = k. \qquad (2)$$

Moreover, by the expansion theorem,

$$\sum_k A_k\phi = \Sigma\alpha_k(\alpha_k, \phi) = \phi,$$

for any vector ϕ. Hence

$$A_1 + A_2 + \ldots + A_n = I. \qquad (3)$$

Any set of projective operators $\{A_j\}$ which satisfies these three conditions is called a "spectral set" for reasons which are explained in Chapter II (pp. 25, 26, 33).

EXAMPLES :—

(1) The cosine of the angle between two vectors α and β being defined (in magnitude) by the equation

$$(\alpha, \alpha)(\beta, \beta) \cos^2 \theta = (\alpha, \beta)(\beta, \alpha),$$

prove that

$$-1 \leqslant \cos^2 \theta \leqslant +1.$$

If the vectors are parallel, i.e. if $\theta = 0$, prove that $\alpha = c\beta$, where c is an ordinary complex number.

[Let

$$(\alpha + z\beta, \alpha + z\beta) = lz^*z + mz^* + m^*z + n$$
$$= l(z + m/l)(z^* + m^*/l) + n - m^*m/l.$$

Then $ln \geqslant m^*m$, etc. (Generalisation of Schwartz's inequality.)]

(2) An operator R is said to be idempotent if $R^2 = R$. If R and S are idempotent, prove that $I - R$ is idempotent, and show that the necessary and sufficient conditions that $R \pm S$ should be idempotent are that

(*a*) $RS = SR = O$, (*b*) $RS = SR = S$.

(3) An operator R is said to be non-negative if $(\alpha, R\alpha) \geqslant 0$ for all vectors α. Prove that a projective operator P cannot be expressed as the sum of two non-negative symmetric operators, A and B, unless these are numerical multiples of P

[If ν is orthogonal to π, $A\nu + B\nu = 0$, i.e. $(\nu, A\nu) + (\nu, B\nu) = 0$, etc.]

(4) If $A_1, A_2, \ldots A_n$ and $B_1, B_2, \ldots, B_n$ are two spectral sets of projective operators, and if

$$U = \sum_k (\beta_k, \alpha_k)^{-1} B_k A_k, \quad V = \sum_k (\alpha_k, \beta_k)^{-1} A_k B_k,$$

show that

$$\begin{aligned} B_k U &= U A_k, \quad A_k V = V B_k, \\ VU &= I = UV, \\ B_k &= U A_k V, \quad A_k = V B_k U. \end{aligned}$$

(5) If A and B are symmetric operators prove that the necessary and sufficient condition that AB should be symmetric is that A should commute with B.

(6) If iA is symmetric, A is said to be "skew." If A is skew prove that $a_{jk} + a_{kj}^* = 0$, and show that if A and B are symmetric (AB + BA) is symmetric and (AB − BA) is skew.

Proper Vectors and Proper Values.—As in two-dimensional space, α is said to be a proper vector of R with proper value c if

$$R\alpha = c\alpha$$

This vector equation is equivalent to the following n scalar equations,

$$\sum_k r_{jk} a_k = c a_j, \quad (j = 1, 2, \ldots, n)$$

which will be compatible if and only if c is a root of the determinantal equation

$$\Delta(c) \equiv \begin{vmatrix} r_{11} - c & r_{12} & \cdots & r_{1n} \\ r_{21} & r_{22} - c & \cdots & r_{2n} \\ \cdot & \cdot & \cdot & \cdot \\ r_{n1} & r_{n2} & \cdots & r_{nn} - c \end{vmatrix} = 0.$$

The roots of this equation, say r_1, r_2, . . ., r_n, are the proper values of R; and if c is given any one of these values, say r_j, the preceding n equations can be solved for the ratios $a_1 : a_2 : \ldots : a_n$ which specify the direction of the associated proper vector ρ_j.

The following theorem is of fundamental importance for symmetric operators: The necessary and sufficient condition that the proper values of S should be *real* and that the proper vectors of S should be mutually *orthogonal* is that S should be symmetric.

Let S have the proper values $s_1, s_2, \ldots s_n$, and let $\sigma_1, \sigma_2, \ldots \sigma_n$ be the corresponding unit proper vectors. Then

$$(\sigma_j, \sigma_k) = 0 \text{ if } j \neq k$$
$$= 1 \text{ if } j = k.$$

To prove the necessity of the condition, let any vector α be expressed in the form

$$\alpha = \sum_j \sigma_j c_j,$$

then

$$S\alpha = \sum_j \sigma_j s_j c_j,$$

and

$$(\alpha, S\alpha) = \sum_{j,k} c_k{}^*(\sigma_k, \sigma_j) s_j c_j = \sum_j c_j{}^* c_j s_j,$$

which is real. Hence, by definition, S is symmetric.

To show that the condition is sufficient, we note that if S is symmetric, then

$$s_j(\sigma_k, \sigma_j) = (\sigma_k, S\sigma_j) = (S\sigma_k, \sigma_j) = s_k{}^*(\sigma_k, \sigma_j).$$

Hence, if $k = j$, $s_j = s_j{}^*$, i.e. s_j is real, and if $k \neq j$,

$$(s^*_k - s_j)(\sigma_k, \sigma_j) = 0, \text{ i.e. } (\sigma_k, \sigma_j) = 0,$$

and σ_k, σ_j are orthogonal (unless $s_k = s_j$).

EXAMPLES :—

(1) If $\sigma_1, \sigma_2, \ldots, \sigma_m$ are proper vectors of S corresponding to an m-times repeated root, s, of $\Delta(c) = 0$, show that the vectors $\tau_1, \tau_2, \ldots, \tau_m$, defined by the recurrence relations,

$$\tau_1 = \sigma_1, \ \tau_2 = \sigma_2 - \tau_1(\tau_1, \sigma_2), \ldots,$$

$$\tau_{k+1} = \sigma_{k+1} - \sum_{j=1}^{k} \tau_j(\tau_j, \sigma_{k+1}), \ldots, (k=2, 3, \ldots, m-1),$$

are mutually orthogonal proper vectors of S for the same proper value s. (This completes the preceding theorem.)

(2) If j is a positive integer, $k = 1, 2, \ldots, 2j + 1$, and

$$s_{k, k-1} = \tfrac{1}{2}(k-1)^{\frac{1}{2}}(2j-k+2)^{\frac{1}{2}} = s_{k-1, k},$$

all the other matrix elements of S being zero, the proper values of S are

$$-j, -j+1, -j+2, \ldots, j-2, j-1, j.$$

Verify for $j = 1, 2, 3$.

[These are the matrix elements of the momentum operator M_x, see p. 56.]

(3) The bare fact of the existence of proper vectors and proper values enables us to give a short and simple proof of the following theorem due to J. v. Neumann :—

If the symmetric operator R *can be expressed as the sum of two non-negative symmetric operators,* A *and* B, *when and only when* A=cR, B=$(1-c)$R, *then* R *is a numerical multiple of a projective operator.*

[Let ρ be a unit proper vector of R with proper value r. Since R must be non-negative, therefore $r \geqslant 0$. Define A, B, and σ by the equations

$$A\phi = r\rho(\rho, \phi), \quad B\phi = R\phi - A\phi,$$
$$\sigma = \phi - \rho(\rho, \phi), \ (\therefore (\rho, \sigma) = 0).$$

Then

$$(\phi, A\phi) \geqslant 0 \quad \text{and} \quad (\phi, R\phi) = (\sigma, R\sigma) + (\phi, A\phi).$$

Hence $(\phi, B\phi) = (\sigma, R\sigma) \geqslant 0$, i.e. A and B are non-negative. Therefore, by hypothesis, A = cR ; but r^{-1}A is projective, etc.]

The Canonical Form of a Symmetric Linear Operator. —Any symmetric linear operator R can be expressed in a very simple form by means of the spectral set of projective operators, R_1, R_2, . . . , R_n, corresponding to the proper vectors of R, $\rho_1, \rho_2, \ldots, \rho_n$. This canonical expression for R is

$$R = r_1R_1 + r_2R_2 + \ldots + r_nR_n,$$

the r_j's being the proper values of R.

To establish this result, denote the right-hand side of the last equation by T. Since

$$R_j\rho_k = 0 \text{ if } j \neq k, \quad \text{or} \quad \rho_k \text{ if } j = k,$$

it follows that

$$T\rho_k = r_k\rho_k = R\rho_k.$$

Hence, if α is any vector,

$$\begin{aligned} T\alpha &= \sum_k T\rho_k(\rho_k, \alpha) = \sum_k R\rho_k(\rho_k, \alpha) \\ &= R\alpha, \end{aligned}$$

i.e. T is identical with R.

It now follows at once that

$$R^k = \sum_j r_j{}^k R_j \quad \text{and} \quad f(R) = \sum_j f(r_j)R_j,$$

where k is any positive integer and $f(R)$ any polynomial in R. Also, if none of the proper values of R are equal to zero, we can define the reciprocal of R, which is written as R^{-1}, and any negative integral power of R, R^{-k}, by the equations

$$R^{-k} = \sum_j r_j{}^{-k}R_j, \quad k = 1, 2, \ldots.$$

It is clear that

$$R^{-1} . R = I = R . R^{-1}, \text{ etc.},$$

so that the definition and notation are justified. It will be noted that projective operators have no reciprocal.

Furthermore, if $f(x)$ is a series of powers of x which is convergent for all values of x, $f(R)$ can be defined by the equation

$$f(R) = \sum_j f(r_j)R_j.$$

To justify this equation, let $f_k(x)$ denote the sum of the first k terms of $f(x)$. Then

$$f(x) - f_k(x) \to 0 \text{ as } k \to \infty\,;$$

and if α is any vector,

$$\begin{aligned} f(\mathrm{R})\alpha - f_k(\mathrm{R})\alpha &= \sum_j [f(r_j) - f_k(r_j)]\mathrm{R}_j\alpha, \\ &\to 0 \text{ as } k \to \infty. \end{aligned}$$

In particular,

$$\begin{aligned} \sum_j \exp(cr_j)\mathrm{R}_j &= \exp(c\mathrm{R}) \\ &= \mathrm{I} + c\mathrm{R} + c^2\mathrm{R}^2/2\,! + \dots \\ &\qquad + c^k\mathrm{R}^k/k\,! + \dots. \end{aligned}$$

Unitary Operators.—If $\{\alpha_j\}$ and $\{\beta_j\}$ are two vector bases, the operator U which transforms any vector of one basis α_j into the corresponding vector of the other basis β_j, is called a "unitary operator." In symbols

$$\beta_j = \mathrm{U}\alpha_j,$$

whence, if ϕ is any vector,

$$\mathrm{U}\phi = \mathrm{U}\sum_j \alpha_j(\alpha_j, \phi) = \sum_j \beta_j(\alpha_j, \phi).$$

Every unitary operator U has a reciprocal U^{-1} defined by the equations

$$\alpha_j = \mathrm{U}^{-1}\beta_j, \quad \mathrm{U}^{-1}\phi = \sum_j \alpha_j(\beta_j, \phi),$$

whence $\mathrm{U}^{-1}\mathrm{U} = \mathrm{I} = \mathrm{U}\mathrm{U}^{-1}$. An explicit expression for U is given in example 4 of page 12.

It follows from the preceding expressions for $\mathrm{U}\phi$ and $\mathrm{U}^{-1}\phi$ that $(\phi, \mathrm{U}\phi)$ and $(\phi, \mathrm{U}^{-1}\phi)$ are conjugate complex numbers, so that in general U is not a symmetric operator. To discuss the proper values and proper vectors of U, let

$$\mathrm{U}\psi = c\psi, \quad \text{and} \quad (\psi, \psi) = 1.$$

Then
$$\mathrm{U}^{-1}\psi = c^{-1}\psi,$$

and
$$(\psi, \mathrm{U}\psi) = c, \quad (\psi, \mathrm{U}^{-1}\psi) = c^{-1}.$$

Hence c and c^{-1} are conjugate complex numbers, i.e. the proper values of U are complex numbers with modulus unity, of the form exp $(i\theta)$, where θ is real. Again, since $U\psi = \sum_k \beta_k(\alpha_k, \psi)$ it follows that

$$(U\psi, U\phi) = \sum_{j,k} (\psi, \alpha_k)(\beta_k, \beta_j)(\alpha_j, \phi)$$
$$= \sum_j (\psi, \alpha_j)(\alpha_j, \phi) = (\psi, \phi).$$

Hence, if σ and τ are any two vectors,

$$(U^{-1}\sigma, \tau) = (\sigma, U\tau).$$

In particular, if σ and τ are proper vectors of U with proper values s and t, $(s^{-1})^*(\sigma, \tau) = t(\sigma, \tau)$. Hence σ and τ are orthogonal unless $s = t$.

If two or more proper values of U are equal we can use the method of Example 1, page 13, to construct a set of proper vectors which shall be mutually orthogonal. Hence in any case it is true that U possesses n mutually orthogonal proper vectors, $\psi_1, \psi_2, \ldots, \psi_n$. If the corresponding projective operators are $H_1, H_2, \ldots, H_n$, it can be proved, as for a symmetric operator, that U can be expressed in the canonical form

$$U = \sum_k \exp(\iota\theta_k) \,.\, H_k,$$

Hence, if H denotes the symmetric operator,

$$H = \sum \theta_k H_k,$$

U can also be expressed in the form $U = \exp(\iota H)$.

Groups of Unitary Operators.—The set of unitary operators $\{U(s)\}$, defined by the equation

$$U(s) = \sum_j \exp(is\theta_j)H_j, \quad (-\infty < s < +\infty),$$

is called a "group" because it satisfies the following two conditions: (*a*) if $U(s)$ belongs to the set so also does $U(-s)$, and $U(s) \,.\, U(-s) = I$; (*b*) if $U(s_1)$ and

$U(s_2)$ belong to the set so also does $U(s_1 + s_2)$ and $U(s_1)U(s_2) = U(s_1 + s_2)$. Such groups of unitary operators are of great importance in the physical theory developed later in Chapter III.

Let $\{\alpha_j\}$ denote some fixed vector basis, and let $\alpha_j(s) = U(s)\alpha_j$. The initial rate of change of $\alpha_j(s)$ with respect to the parameter s, say $\dot{\alpha}_j$, is clearly

$$\lim_{s \to 0} s^{-1}[\alpha_j(s) - \alpha_j]$$

and, since $U(s)$ can be expressed in the form

$$U(s) = I + isH + (isH)^2/2\,! + \ldots + (isH)^k/k\,! + \ldots,$$

it follows that $\dot{\alpha}_j = iH\alpha_j$. For this reason, the operator H is called the " infinitesimal operator " of the group $\{U(s)\}$.

So far we have studied only the transformation of *vectors* by unitary operators. It is now necessary to study the transformation of operators. An operator A can be specified by its matrix elements $a_{jk} = (\alpha_j, A\alpha_k)$ relative to the fixed basis $\{\alpha_j\}$. The operator B, which corresponds to A in the vector basis $\{\beta_j\}$, is defined by the equations

$$(\beta_j, B\beta_k) = (\alpha_j, A\alpha_k).$$

Hence, if $\beta_j = U\alpha_j$, it follows that

$$(\beta_j, B\beta_k) = (U^{-1}\beta_j, AU^{-1}\beta_k) = (\beta_j, UAU^{-1}\beta_k),$$

i.e.

$$B = UAU^{-1}.$$

Similarly, if $A(s)$ in the basis $\{\alpha_j(s)\}$ corresponds to A in the fixed basis $\{\alpha_j\}$, then

$$A(s) = U(s)AU(-s),$$

or

$$U(s)A(s) = U(s)A.$$

The initial rate of change of $A(s)$ with respect to s, say, $\dot{A}$, is defined by the equation

$$\dot{A} = \lim_{s \to 0} s^{-1}[A(s) - A],$$

and it easily follows that

$$\dot{A} = \iota(HA - AH)$$

where H is the infinitesimal operator of U.

EXAMPLES :—

(1) If $(U\alpha, U\alpha) = (\alpha, \alpha)$ for all vectors α, prove that U is unitary.

(2) If S is a symmetric operator, prove that

$$U \equiv c(S - iI)(S + iI)^{-1}$$

is unitary ($c^{-1} = c^*$); and if U is unitary, prove that

$$S \equiv i(cI + U)(cI - U)^{-1}$$

is symmetric.

(3) If $\{\epsilon_j\}$ is a basis and

$$U\epsilon_j = \epsilon_{j+1}, \ (j = 1, 2, \ldots, n - 1), \ U\epsilon_n = \epsilon_1,$$

prove that U is unitary and find its proper values and vectors. Construct the matrices for U, $U^2 \ldots U^{n-1}$ and prove that $U^n = I$.

(4) Show that the matrix

$$\left\| \begin{matrix} \cos s\theta, & -\sin s\theta \\ \sin s\theta, & \cos s\theta \end{matrix} \right\|$$

specifies a group of unitary operators $\{U(s)\}$, and find the infinitesimal operator of this group. [It is the operator J of p. 2]

(5) A three-dimensional representation of the angular momentum operators M_x, M_y, M_z, of page 49 is provided by the matrices

$$-ih\left\| \begin{matrix} 0 & 0 & 0 \\ 0 & 0 & +1 \\ 0 & -1 & 0 \end{matrix} \right\|, \ -ih\left\| \begin{matrix} 0 & 0 & -1 \\ 0 & 0 & 0 \\ +1 & 0 & 0 \end{matrix} \right\|, \ -ih\left\| \begin{matrix} 0 & +1 & 0 \\ -1 & 0 & 0 \\ 0 & 0 & 0 \end{matrix} \right\|.$$

Find the proper values and proper vectors of M_z, and determine the unitary operator U which transforms the given basis into these proper vectors. Express M_x and M_y in terms of this new basis.

The Characteristic of an Operator.—The "characteristic" or "trace" of an operator R is defined to be the sum of its proper values, $r_1 + r_2 + \ldots + r_n$, and is denoted by the symbol $\chi(R)$. Since the proper values are the roots of the determinantal equation, $\Delta(c) = 0$, (p. 12), it follows that

$$\chi(R) = r_{11} + r_{22} + \ldots + r_{nn},$$

i.e. that the characteristic of R is the sum of the diagonal elements of the matrix representing R.

Now the values of the matrix elements of R, which are given by the formula,

$$r_{jk} = (\epsilon_j, R\epsilon_k),$$

obviously depend upon the basis, $\epsilon_1, \epsilon_2, \ldots, \epsilon_n$. If $\epsilon_1', \epsilon_2', \ldots, \epsilon_n'$ is another basis, obtained from the original basis by the unitary transformation

$$\epsilon_j' = U\epsilon_j, \quad (j = 1, 2, \ldots, n),$$

the matrix elements of R in the new basis will be

$$r_{jk}' = (\epsilon_j', R\epsilon_k') = (U\epsilon_j, RU\epsilon_k) = (\epsilon_j, U^{-1}RU\epsilon_k),$$

i.e. equal to the matrix elements of $U^{-1}RU$ in the original basis. Hence, if u_{jk} are the matrix elements of U in the original basis, then

$$r_{jk}' = \sum_{s,t} u_{sj}^* r_{st} u_{tk}.$$

Since the characteristic of an operator has been defined in terms of its intrinsic properties, it follows that it is invariant for any unitary transformation U, i.e.

$$\chi(R) = \chi(U^{-1}RU).$$

EXAMPLES :—

(1) Prove that $\chi(AB) = \chi(BA)$ for any two operators A and B.

(2) Show that $\chi(U)$ and $\chi(U^{-1})$ are conjugate complex numbers for a unitary operator U.

(3) If the N operators $P_1 = I, P_2, \ldots P_N$ form a group, prove that

$$\chi(P_j)I = \sum_k P_k P_j P_k^{-1}.$$

(A group of operators is defined by the conditions that it contains (*a*) the identical operator, (*b*) the inverse of any operator, (*c*) the product of any two operators.)

CHAPTER II

THE LAWS OF MEASUREMENT IN ATOMIC PHYSICS

THE foundations of the quantum theory are the laws of measurement in atomic physics, i.e. the general principles of the interpretation of experimental observations made on the ultimate elements of the physical world. The surprising conclusion to be drawn from these principles is that the physical characteristics of the ultimate elements are appropriately represented in mathematical language *not* by ordinary numbers but by *linear operators* of the type studied in the first chapter.

Atomic Physics.—Macroscopic physics and microscopic physics differ widely in their formal objects, their methods and concepts. Macroscopic physics studies the properties of gross matter, in bulk large enough to be directly observed by the senses. Its standards of measurement are measuring rods, clocks and weights with a direct appeal to hand and eye. Its extreme accuracy of observation is due to the possibility of successively subdividing gross matter in finer and finer divisions which are still distinguishable by eye or microscope. Macrophysical experiments are made on individual physical objects, and the results of these experiments are summarised in definite numerical values with small and known ranges of proportionate error.

But microphysics studies the ultimate, or penultimate, elements of matter—such as the electron and proton, the atom and molecule. In this domain direct observation by the senses is useless. The existence and properties of the ultimate elements are only to be inferred indirectly from observations of gross matter, e.g. by observations of the action of individual electrons on oil-drops as in Millikan's experiments, or observations of large aggregates of molecules as in the study of molecular rays. The outstanding inference to be drawn from such observations is the existence of certain natural, indivisible units,—the various material units of mass such as the electron and proton, the electrical unit of charge, e, and the dynamical units related to Planck's unit of action, h. It appears that the emission and absorption of energy, linear momentum and angular momentum in the form of light of frequency ν and wavelength λ take place in units of magnitude $h\nu$, h/λ and h, and that the energy and momentum of stable atomic and molecular structures do not vary continuously but change by discontinuous jumps. This material, electrical and dynamical atomicity characteristic of microphysics makes the problem of microphysical measurement overwhelmingly difficult.

The Problem of Microphysical Measurement.—All measurement involves the paradox that the system measured must be at the same time an isolated whole and a part interacting with other parts. Measurement is impossible unless the system acts upon the apparatus of observation, and the measurements are meaningless unless the system retains its identity and characteristics. The action of the system on the apparatus is always accompanied by a reaction of the apparatus on the system, which thereby suffers some change of state. In macrophysics the interaction of the system and the apparatus can always be made negligibly small but in microphysics the atomicity of energy, momentum and action places a natural lower limit on the magnitude of this interaction—and a limit which is of the same order of magnitude as the quantities to be measured. Hence microphysical experiments may produce large and non-negligible changes in the systems measured.

Under such circumstances the most satisfactory type of measurement which can be made is one which leaves unchanged that characteristic which it is designed to measure, although it may perhaps produce changes in other characteristics. Thus the velocity of a stream of electrons can be measured by opposing to its motion a potential barrier or by deflecting it by a magnetic field. The first method destroys the velocity which is measured, but the second method preserves the velocity unchanged, although it may alter other characteristics, such as the orientation of the magnetic axes of the electrons. Only measurements of the second kind are really entitled to be called "observations" in microphysics, and henceforward this word will be used in this special sense.

The Law of Simple Observation.—An observation of a characteristic of a system is therefore a measurement of this characteristic which produces no change in its value. Such a definition needs to be completed by the specification of an experimental test to decide which measurements are "observations" within the meaning of the definition. Obviously this criterion for an observation is that its immediate repetition shall yield again the same result as the first experiment.

In developing this criterion it must be borne in mind that microphysical experiments are made upon large assemblies of systems, in which the magnitude of the characteristic to be measured may be continuously distributed over a wide range, or may be confined to certain discrete values. Thus the kinetic energies of silver atoms (in the normal state) emerging from a furnace may have any value, but their velocities will not exceed the velocity of light, and their magnetic moments will be $\pm$ 1 Bohr magneton. A general theory must take into account all these possibilities. In dealing with any variable y it is accordingly necessary not to prejudice the question by speaking of "an observation that y has the value a," for, if the possible values of y are distributed discretely, a may not be one of them, and, even if the possible values are continuously distributed, only an infinitesimal (i.e. unobservable) proportion of systems could have a value of y exactly equal to a.

Accordingly the simplest general type of observation is that the value of a variable y lies within a certain range, $y_n \leqslant y < y_{n+1}$, where the y_n's are arbitrary, convenient numbers ranging from $-\infty$ to $+\infty$. Such an observation will be called a "simple observation," and will be denoted by the symbol Y_n.

The symbol Y_n represents the physical process of observing that $y_n \leqslant y < y_{n+1}$ for certain systems in a given assembly. This implies that the process can discriminate between systems for which y lies in different ranges, $y_j \leqslant y < y_{j+1}$ and $y_k \leqslant y < y_{k+1}$, and this in turn implies that the process produces a real or virtual separation of the systems so characterised. If α denotes the given assembly and η_n the set of systems for which $y_n \leqslant y < y_{n+1}$, the physical process of separation is conveniently represented by the symbolic equation

$$Y_n\alpha = \eta_n.$$

The condition that the process should be a "simple observation" is then represented by the equation

$$Y_n\eta_n = \eta_n,$$

i.e. $$Y_nY_n\alpha = Y_n\alpha,$$

for any given assembly α. Since the symbol α is clearly irrelevant, the last equation may be written

$$Y_nY_n = Y_n \quad \text{or} \quad Y_n^2 = Y_n, \qquad (1)$$

i.e. Y_n is an idempotent operator. This equation symbolises the "law of simple observation."

The Laws of Complex Observation.—The whole process of measuring the variable y for a given assembly α requires the determination of the systems for which y lies in each of the ranges,

$$\ldots, y_1 \leqslant y < y_2,\ y_2 \leqslant y < y_3, \ldots, \\ y_n \leqslant y < y_{n+1}, \ldots$$

The corresponding physical process will be called a "complex observation." It is represented by the set of operators $Y_1, Y_2, \ldots Y_n, \ldots$ which, applied to the given mixed assembly α, yield the " pure " assemblies

$$\ldots \eta_1, \eta_2, \ldots, \eta_n, \ldots$$

This process of separation must satisfy the two obvious conditions of " exclusiveness " and " completeness."

First, the simple observation represented by Y_k separates out systems for which $y_k \leqslant y < y_{k+1}$. Hence, if this observation is applied to the assembly η_j, for which $y_j \leqslant y < y_{j+1}$, *no* systems will be separated from η_j in the way in which the process Y_k usually separates systems. This property of the Y_n's is symbolised by the equation

$$Y_k \eta_j = o,$$

i.e.
$$Y_k Y_j \alpha = o \quad \text{or} \quad Y_k Y_j = O, \qquad (2)$$

where O is the " nul operator," which separates no systems from any given assembly.

Secondly, every system in the given mixed assembly α must pass into one of the pure assemblies, $\eta_1, \eta_2, \ldots \eta_n \ldots$, which are separated from α by the set of simple observations which together make up the complex observation. This property of the Y_n's may be symbolised by the equation

$$\sum_n \eta_n = \alpha,$$

i.e.
$$\Sigma Y_n \alpha = \alpha,$$

or
$$\sum_n Y_n = I, \qquad (3)$$

where I is the " identical operator," which separates *all* the systems from any given assembly.

Equations (2) and (3) symbolise the " laws of complex observation."

Compatible and Incompatible Observations. — In general an " observation " which is designed to measure the distribution of the values of a variable y in a given assembly will produce changes in the distribution of the values of another variable z, but, in exceptional cases, it may happen that no such interference takes place. In the first case the variable y and z are said to be " incompatible," and in the second case they are said to be " compatible," the same adjectives being applied to the corresponding observations. These definitions can be expressed symbolically in terms of the symbols representing " simple observations " of y and z.

As before let Y_j represent the simple observation which separates systems for which $y_j \leqslant y < y_{j+1}$, and let Z_k represent the simple observation which separates systems for which $z_k \leqslant z < z_{k+1}$. If the variables y and z are compatible the compound observations, Y_jZ_k or Z_kY_j consisting of Z_k followed by Y_k, or *vice versa*, will both yield an assembly for which $y_j \leqslant y < y_{j+1}$ and $z_k \leqslant z < z_{k+1}$, i.e.

$$Y_jZ_k = Z_kY_j.$$

But, if the variables are incompatible, the second observation will disturb the characteristic selected by the first observation, and, under these circumstances,

$$Y_jZ_k \neq Z_kY_j.$$

The maximum of information regarding a given assembly will clearly be obtained by employing the maximum number of compatible observations. For the pure assemblies separated by this means the values of the maximum number of variables will be known to fall within definite ranges. It will then be impossible to obtain information about the values of any other variables, as the attempt to do so would produce changes in the values of the variables already measured. The pure assembly (" dass reine Fall " (Weyl), " die einheitliche Gesamtheit " (v. Neumann)) prepared by the operation of a maximum number of simple observations has therefore the greatest definiteness which can be expected in microphysics. The values, or rather the ranges, of the variables which define such an assembly specify the *state* of the systems in the assembly. The state of an assembly is also determined by the set of compatible simple observations, Y_j, Z_k, W_l . . . ., by which it has been separated. The complete observation is represented by the operator

$$P = Y_jZ_kW_l \; . \;\; . = Z_kY_jW_l \; . \; . \; . = . \; . \; .,$$

which obviously satisfies the condition that

$$P^2 = P.$$

$$(\text{for } P^2 = Y_jZ_kW_l \; . \; . \; . \; Y_jZ_kW_l = Y_j{}^2Z_k{}^2W_l{}^2 \; . \; . \; .$$
$$= Y_jZ_kW_l \; . \; . \; . = P).$$

The Laws of Transitions.—Let α represent an assembly in a definite state for which $y_j \leqslant y < y_{j+1}$, $z_k \leqslant z < z_{k+1}$, etc. If l is not one of the variables which define the state of α it is impossible to measure the value of l for systems in α, as any attempt to do so will produce changes in the values of y, z, etc. Of course, if the assembly α is subjected to the simple observation L_n, there will be separated an assembly λ_n, and for the systems in this assembly, $l_n \leqslant l < l_{n+1}$. Similarly, the assembly α may be subjected to a complete observation Q which yields another assembly β in a definite state: The observation Q will have produced a number of changes in the systems which composed the assembly α and which now compose the assembly β. These systems have undergone a transition from the state of the assembly α to the state of the assembly β, and the simplest assumptions regarding such transitions are the following :—

1. There is definite probability p that a system in the state specified by α or by P will pass over into the state specified by β or by Q. This probability may be written as $p(\alpha \rightarrow \beta)$ or $p(P \rightarrow Q)$.

2. The probability of a transition connecting the states specified by α and by β is independent of the *direction* of the transition, i.e. $p(\alpha \rightarrow \beta) = p(\beta \rightarrow \alpha)$.

Without the first assumption it would be impossible to make any further progress in quantum theory. It represents the irreducible minimum required for the investigation of causal relations in microphysics. It corresponds to the fundamental assumption of macrophysics that a *complete* knowledge of the present state of a system furnishes sufficient data to determinate definitely its state at any future time or its response to any external influence. The second assumption, for which there is no direct evidence, is an extension of the Principle of Detailed Balancing as used in Statistical Mechanics. It appears to correspond to the reversibility of processes in macrophysics. The two assumptions will be taken to be the general laws regulating transitions (American " switches "), and $p(\alpha \rightarrow \beta)$ will be called the " transition probability " from state α to state β.

To obtain the symbolical expression of these laws consider the series of three complete observations represented by PQP. The final result of this triple process will be an assembly in the same state as the assembly separated by the single process P. Hence, α being the initial mixed assembly, the assemblies $\mathrm{PQP}\alpha$ and $\mathrm{P}\alpha$ can differ only in the number of systems which they contain, and we shall therefore write

$$\mathrm{PQP}\alpha = c\mathrm{P}\alpha \quad \text{or} \quad \mathrm{PQP} = c\mathrm{P},$$

where c is some proper fraction. Then by the second transition law we shall also have

$$\mathrm{QPQ} = c\mathrm{Q},$$

with the same constant c in both cases. This constant will later (p. 31) be identified with the transition probability $p(\mathrm{P} \rightarrow \mathrm{Q}) = \mathrm{p}(\mathrm{Q} \rightarrow \mathrm{P})$.

The Representation of States by Vectors.—The preceding sections of this chapter summarise the fundamental laws of measurement in atomic physics. In this and the succeeding sections those laws will be used to discuss the mathematical representation of " simple " and " complex " observations, and of the variables characterising physical systems.

The mathematical representation of simple observations is suggested by the following result : If $\eta_1, \eta_2, \ldots$

are the "pure" assemblies separated from a given assembly α in a definite state by the complex observation represented by Y_1, Y_2, . . ., and if p_1, p_2, . . . are the probabilities of a transition from α to η_1, η_2, . . ., respectively, then, clearly,

$$p_1 + p_2 + \ldots = 1,$$

for every system in α must pass into one of the assemblies η_1, η_2, ... Now the transition probabilities, p_1 p_2, . . ., describe the relation of α to η_1, η_2, ... Hence this relation can be stated in geometrical language by saying that (i) α can be represented by a unit vector, (ii) η_1, η_2, . . . can be represented by a set of orthogonal vectors, and (iii) the components of α in the direction of η_1, η_2, . . . are numbers a_1, a_2, . . . such that

$$|a_k|^2 = p_k \ (k = 1, 2, \ldots).$$

This suggests that it should be possible to represent any assembly of systems in a definite state by a vector. Now a vector has both magnitude and direction, and presumably two vectors in the same direction but with different magnitudes would represent two assemblies of systems in the *same* state, but containing different numbers of systems. Also it is to be anticipated that the divergence in direction of two vectors will be associated with the probability of a transition between the assemblies which they represent.

In order to examine the possibility of such a representation we begin by agreeing to represent the *states* of the pure assemblies η_1, η_2, . . ., separated from α by Y_1, Y_2, . . ., by a set of *unit*, orthogonal vectors ϵ_1, ϵ_2, . . ., the actual assemblies themselves being represented by vectors $c_1\epsilon_1$, $c_2\epsilon_2$ . . . when c_1, c_2 . . . are certain numbers. Similarly, we agree to represent the given assembly α by some *unit* vector α with components a_1, a_2, . . . referred to the vectors ϵ_1, ϵ_2, . . . as a basis. We shall use the same symbols for vectors and for the assemblies which they represent.

So far our representation is a pure convention. The

processes Y_1, Y_2, . . . must be represented by operators (which we shall denote by the same symbols) such that

$$Y_k\alpha = c_k\epsilon_k,$$
$$Y_k{}^2\alpha = Y_k\alpha,$$
and
$$Y_jY_k\alpha = 0. \qquad (j \neq k.)$$

Hence Y_1, Y_2, . . . will be the projective operators (in the sense of Chap. I) associated with the unit vectors ϵ_1, ϵ_2, ... Moreover, we must have

$$a_k = c_k,$$

and $|a_k|^2$ = probability of the transition $\alpha \rightarrow \eta_k$, i.e. the number of systems in the assembly η_k is proportional to the square of the magnitude of the vector η_k. This must be true for every assembly.

We can now identify the constant c in the second transition law,

$$ABA = cA,$$

where A, B, represent simple observations. In each process A $\rightarrow$ B and B $\rightarrow$ A the number of systems is reduced in the ratio $1 : p$, where p is the transition probability common to both processes. Hence the double process A $\rightarrow$ B, B $\rightarrow$ A reduces the number of systems in the ratio $1 : p^2$. Hence, if the original assembly is represented by that unit vector α, the final assembly will be represented by the vector $p\alpha$. Accordingly, if $A\phi = \alpha$, then $ABA\phi = p\alpha$, whence $c = p$, i.e. c is the transition probability $p(A \rightarrow B)$.

Now it was shown in Chapter I that if A and B are projective operators the constant c, defined as above, is simply $|(\alpha, \beta)|^2$, where α and β are the unit vectors associated with A and B. Hence we have shown that, as a consequence of the laws of measurement in atomic physics, *it is possible to represent the states of systems by unit vectors*, α, β, . . . , *in such a way that the transition probabilities* $p(\alpha \rightarrow \beta)$ *are equal to the squared moduli of the scalar products* $|(\alpha, \beta)|^2$.

A simple example is given by Dirac (" The Principles of Quantum Mechanics," § 2). Consider the action of a polariscope on a beam of plane polarised light. The incident light (which here corresponds to the original assembly) is separated into two components which are polarised at angles θ and $\theta + \frac{1}{2}\pi$ radians with the direction of polarisation of the original beam, and the intensities of this and of the two emergent beams are in the ratio

$$1 : \cos^2\theta : \sin^2\theta = \cos^2(\theta + \tfrac{1}{2}\pi).$$

Hence, if the state of polarisation of the incident beam is represented by a unit vector, the states of the emergent beams may be represented by unit vectors making angles θ and $\theta + \frac{1}{2}\pi$ with the first vector, and the transition probabilities will be represented by the squared cosine of the angle between the appropriate vectors.

The Stern-Gerlach Experiment.—In the realm of atomic physics itself the most simple and direct experimental illustrations of the "laws of measurement" enunciated above are derived from observations of *molecular rays*. One of the earliest and most striking experiments of this type is that devised by Stern and Gerlach for the measurement of the magnetic moments of metallic atoms. The metal is vaporised in a furnace, and the issuing stream of atoms is so limited by slits that it emerges as a narrow pencil. This "ray" of atoms passes across an inhomogeneous magnetic field and is then allowed to impinge on a screen so as to yield an observable trace.

It is clear that those atoms which possess a magnetic moment will be deviated from the original direction of the pencil, the deviation being in the direction of magnetic field, and that the amount of this deviation will depend upon the component of their magnetic moment parallel to the field.* Hence the traces on the screen form a "spectrum" of the atoms with reference to this characteristic. It is found that the spectrum is not continuous but discrete, and that it consists of a few

* A brief account of the theory is given by E. C. Stoner in "**Magnetism,**" p. 12 (Methuen's Monographs on Physical Subjects, No. 5).

fairly sharp lines, corresponding to the possible values of the field component of the atomic magnetic moments. This is a simple example of the resolution of a given aggregate of atoms into a number of " pure " aggregates in each of which the measured characteristic (the magnetic moment) has a definite proper value.

The original Stern-Gerlach experiment is simply the spectral analysis of a given inhomogeneous aggregate of atoms. It would be an experiment of extreme interest to repeat this process of spectral analysis on one of the pure aggregates separated out by the first experiment, especially if the general direction of the magnetic field were made different in the second experiment. With such an arrangement one could verify the general laws of transitions adopted above, and the " exchange relations " obtained in a later chapter. A number of such experiments are being made by the Hamburg school, and the results are in fairly satisfactory agreement with theory.

The " Spectrum " of a Variable.—It will be clear from the general considerations developed above and from the particular instance of the Stern-Gerlach experiment that a " complete " observation is a species of spectral analysis in which a given inhomogeneous aggregate α is resolved into a number of parts which are (relatively) homogeneous with respect to some variable y. The " spectrum " of a given assembly α may be precisely defined as follows :—

Let $p(x)$ denote the fraction of the total number of systems in α for which $y < x$. Then $p(x)$ is a monotone increasing function of x such that $0 \leqslant p(x) \leqslant 1$. If $p(x)$ is discontinuous when $x = \xi$, then ξ is a point of the discrete spectrum of α and the discontinuity $p(x)$ at ξ measures the fraction of systems in α for which $y = \xi$. But, since α is a statistical assembly containing an infinite number of systems, for some ranges of values of x $p(x)$ may vary continuously and possess a derivative $dp(x)/dx = p'(x)$. If $p(x)$ is continuous when $x = \xi$, and $p'(\xi)$ is not zero, then ξ is a point of the continuous spectrum of α. The remaining points at which $p(x)$ is continuous but $p'(x)$ *is* zero do not form part of the spectrum of α.

This suffices to define the spectrum of *a given assembly* α. The spectrum of the *variable* y can now be defined to be the complete set of values of y which includes the spectrum of *every* assembly α. Those values which form part of the continuous spectrum of *any* assembly compose the continuous spectrum of the variable, and the remaining values of the spectrum of y form the discrete spectrum.

EXAMPLE :—

Show that the number of points in the discrete spectrum of an assembly is (at most) enumerably infinite. [The number of points at which the amount of the discontinuity δ in $p(x)$ lies in the range

$$2^{-n} \leqslant \delta < 2^{-n+1}$$

is not greater than 2^n, and is therefore finite, etc.]

The Representation of Variables by Linear Operators. —The mathematical representation of variables by linear operators is an immediate application of the results obtained above relative to the representation of states by vectors.

Let A_j denote the simple observation (and its representative operator !) which separates systems for which $a_j \leqslant a < a_{j+1}$. The A_j's are projective operators, and they are represented by certain definite matrices relative to any set of unit, orthogonal vectors, $\epsilon_1, \epsilon_2, \ldots$ representing the set of pure states determined by a certain complete observation. Hence the A_j's, regarded as matrices, can be added together and multiplied by ordinary numbers.

Now consider the operator

$$A = a_1A_1 + a_2A_2 + \ldots$$

This is a symmetric linear operator, like the A_j's themselves, and has a definite matrix representation referred to any chosen basis $\epsilon_1, \epsilon_2, \ldots$. It will be shown that *in a certain conventional sense this operator* A *represents the variable a.*

To do this we form the scalar product $(\phi, \mathrm{A}\phi)$, where the unit vector ϕ represents some definite pure state of the system considered.

Let $$\phi = \sum_j \alpha_j f_j,$$

then $$\mathrm{A}\phi = \sum_j a_j \alpha_j f_j,$$

and $$(\phi, \mathrm{A}\phi) = \sum_j f_j^* a_j f_j,$$

since the vectors α_j, which represent the pure states of the aggregates separated by the processes A_j, are unit, orthogonal vectors. Now $f_j^* f_j$ equals the probability p_j of the transition $\phi \to \alpha_j$, i.e. it is the fraction of the systems in the assembly ϕ for which $a_j \leqslant a < a_{j+1}$. Hence, if the average or "expected" value of the variable a for the aggregate ϕ is $\mathrm{E}_\phi(a)$,

then $$\sum_j p_j a_j \leqslant \mathrm{E}_\phi(a) < \sum_j p_j a_{j+1}.$$

Hence $(\phi, \mathrm{A}\phi)$ is approximately equal to $\mathrm{E}_\phi(a)$, the absolute error being less than

$$\Sigma p_j(a_{j+1} - a_j),$$

and therefore less than $\sum_j p_j \delta = \delta$, where δ is the maximum value of the ranges $a_{j+1} - a_j$. If the variable a has no continuous spectrum, and if the values a_j are taken to be the proper values forming the discrete spectrum, then we have the accurate result

$$\mathrm{E}_\phi(a) = (\phi, \mathrm{A}\phi),$$

but if a has a continuous spectrum this result is only approximate. The error involved can be made as small as we please by sufficiently decreasing δ, but complete accuracy requires a limiting process in which the sum $\sum_j p_j a_j$ is replaced by an integral. This development will not be required in this book.

The Uniqueness of the Representation of Variables by Operators.—It has been shown that if the variable a has a discrete spectrum only, say at the points a_1, a_2, . . . , then the symmetric linear operator $A = \sum_j a_j A_j$ represents the variable a in the sense that $E_\phi(a) = (\phi, A\phi)$, for any assembly ϕ. It is easily proved that this representation is unique, i.e. that A is the only linear operator for which this result is universally true.

For, if B were another possible representation of a, then $E_\phi(a) = (\phi, B\phi)$, whence $(\phi, R\phi) = 0$ for all ϕ, R being the difference A − B. Let ϵ_1, ϵ_2, . . . be a vector basis. If $\phi = \epsilon_k$, it follows that the diagonal matrix element r_{kk} is zero. If $\phi = \epsilon_j f_j + \epsilon_k f_k$, it follows that

$$r_{jk} \exp(i\theta) + r_{kj} \exp(-i\theta) = 0,$$

where θ is the amplitude (argument) of $f_j^* f_k$. Hence, by putting $\theta = 0$ or $\frac{1}{2}\pi$, we obtain the equations

$$r_{jk} + r_{kj} = 0, \quad i(r_{jk} - r_{kj}) = 0.$$

Therefore all the matrix elements of R are zero, i.e. R = O or B = A, and the representation of a as a linear operator is unique.

The same theorem can be established by a similar argument, even if a has a continuous spectrum, and this result will be assumed in the next section.

The Representation of the Sum or Product of Two Variables.—Let x and y be two variables represented by the symmetric linear operators X and Y. Since

$$E_\phi(x) = (\phi, X\phi) \text{ and } E_\phi(y) = (\phi, Y\phi),$$

it follows that

$$E_\phi(x + y) = E_\phi(x) + E_\phi(y) = (\phi, [X + Y]\,\phi),$$

i.e. the average value of the sum $x + y$ in any state ϕ is obtained by using the sum of the two operators X + Y. Hence $x + y$ is represented by the operator X + Y, and, by the theorem of the preceding section, this representation is unique. Similarly, if c is any ordinary real number, it follows that the operator representing the variable cx is $c\mathrm{X}$. The preceding result is true whether the variables x and y are compatible or incompatible.

To determine the operator representing x^2 we note that if $x_1, x_2, \ldots$ are the proper values of x, then x_1^2, $x_2^2, \ldots$ are the proper values of x^2, while the corresponding projective operators $\mathrm{X}_1, \mathrm{X}_2, \ldots$ are clearly the same as for $x_1, x_2, \ldots$ Hence the operator representing x^2 is

$$x_1^2\mathrm{X}_1 + x_2^2\mathrm{X}_2 + \ldots = (x_1\mathrm{X}_1 + x_2\mathrm{X}_2 + \ldots)^2 = \mathrm{X}^2.$$

Similarly, the operator representing x^n is X^n, and, if $f(x)$ is any polynomial in x, the corresponding operator is $f(\mathrm{X})$. We shall use the notation, $x \to \mathrm{X}$, $x^n \to \mathrm{X}^n$, and $f(x) \to f(\mathrm{X})$ to express concisely the association of a variable and its representative operator.

It is now possible to discuss the representation of the product xy. Since

$$xy = \tfrac{1}{4}(x + y)^2 - \tfrac{1}{4}(x - y)^2,$$

and
$$(x \pm y) \to (\mathrm{X} \pm \mathrm{Y}),$$

it follows by the preceding results of this section that

$$\begin{aligned} xy &\to \tfrac{1}{4}(\mathrm{X} + \mathrm{Y})^2 - \tfrac{1}{4}(\mathrm{X} - \mathrm{Y})^2 \\ &= \tfrac{1}{2}(\mathrm{XY} + \mathrm{YX}), \end{aligned}$$

i.e.
$$xy \to \tfrac{1}{2}(\mathrm{XY} + \mathrm{YX}).$$

In the particular case when x and y are compatible and X and Y commute, this reduces to XY or YX, but in general the operator which represents xy must be expressed as the mean of XY and YX.

The Square Root of an Operator.—It only remains to discuss the representation of the square root of x. If we interpret $x^{\frac{1}{2}}$, as is customary, to mean the *positive* value of the square root of x, then the proper values of $x^{\frac{1}{2}}$ are, with the same convention, $x_1^{\frac{1}{2}}, x_2^{\frac{1}{2}} \ldots.$ Hence, as in the case of x^2 and x^n.

$$x^{\frac{1}{2}} \to x_1^{\frac{1}{2}}\mathrm{X}_1 + x_2^{\frac{1}{2}}\mathrm{X}_2 + \ldots$$

The expression on the right-hand side is *one* of the square roots of X, and this particular root will be denoted in future by $X^{\frac{1}{2}}$ (the other roots are obtained by giving negative signs to some of the $x_k^{\frac{1}{2}}$'s).

EXAMPLES :—

(1) If x and y are compatible, prove that X and Y commute, i.e. XY = YX.
(2) If X and Y are symmetric, prove that X + Y, cX and $\frac{1}{2}$(XY + YX) are symmetric.
(3) If $x, y, z \rightarrow$ X, Y, Z, find the operator representing xyz.
(4) What variable is represented by the projective operator X_k ?

Wave Functions.—An accurate representation of variables with continuous spectra is obtained by a limiting process which we shall now study in detail.

Let the entire range of such a variable be divided up into intervals of equal extent δ, and let $(y, y + \delta)$ be a typical interval. Let the unit vector $\eta(y)$ represent an assembly for which the variable lies in this range. The set of vectors, $\eta(y)$, for all the intervals, clearly forms a "basis." Let $a(y) = (\eta(y), \alpha)$ denote the component of a vector α in the direction of $\eta(y)$. If α is a unit vector, the probability of agreement of α and $\eta(y)$ is $/a(y)/^2$. The component $a(y)$ itself is sometimes described as a "probability amplitude," and it is convenient to introduce here two other concepts: the "probability density," defined as $/a(y)/^2/\delta$, and the "amplitude density," $p(y) = a(y)/\delta^{\frac{1}{2}}$.

On proceeding to the limit the finite interval δ is replaced by the infinitesimal interval dy, and the discrete set of amplitude densities $p(y)$ is replaced by a continuous function $\psi(y) = \lim\limits_{\delta\rightarrow 0} a(y)/(\delta)^{\frac{1}{2}}$. The components of the vector α become the infinitesimals $\psi(y)(dy)^{\frac{1}{2}}$ and the probability density becomes $/\psi(y)/^2$. The function $\psi(y)$ is called the "wave function for the state α," for reasons which will appear later (p. 61).

If $\psi_\alpha(y)$ and $\psi_\beta(y)$ are the wave functions for the states α and β, the scalar product (β, α) is transformed by passage to the limit according to the equations

$$(\beta, \alpha) = \Sigma b^*(y)a(y) = \Sigma p_\beta^*(y)p_\alpha(y)\delta$$
$$\rightarrow \int \psi_\beta^*(y)\psi_\alpha(y)dy,$$

the integral being taken over the entire range of values of the variable. In particular we note that the function $\psi(y)$ is " normalised," i.e.

$$\int |\psi(y)|^2 dy = (\alpha, \alpha) = 1.$$

When the complete specification of the state of an assembly requires the determination of *several* commuting variables with continuous spectra, such as the Cartesian co-ordinates x, y, z, the preceding theory requires extension. The fundamental vectors $\eta(x, y, z)$ now represent the state of systems for which the variables lie in the intervals $(x, x + \delta x\,;\; y, y + \delta y\,;\; z + \delta z)$. The amplitude density $p(x, y, z)$ is now defined as $a(y)/(\delta x \,.\, \delta y \,.\, \delta z)^{\frac{1}{2}}$, $a(y)$ being the component of α in the direction of $\eta(x, y, z)$, and the wave functions $\psi(x, y, z)$ are the limiting values of the amplitude densities. The scalar product (β, α) now becomes the triple integral over all space,

$$\int \psi_\beta^*(x, y, z) \,.\, \psi_\alpha(x, y, z)\, dx\, dy\, dz.$$

CHAPTER III

THE EXCHANGE RELATIONS AND THE EQUATIONS OF MOTION

THE laws of measurement in atomic physics, developed in the preceding chapter, are dominated by the fact that, in general, every microphysical observation produces an unknown and non-negligible change in some of the characteristics of the system observed. Thus a measurement of the values of a variable y may produce changes in the values of another variable z.

The present chapter expresses this principle in a precise, quantitative form by means of Heisenberg's exchange relations (Vertauschungsrelationen), and applies these relations to develop the theory of the angular momentum operators, and to obtain the equations of motion in quantum mechanics.

The General Form of the Principle of Uncertainty.— A microphysical " observation " of the distribution of the values of a variable y in an assembly α allows us to determine the average values of y and y^2, which are denoted by $E(y)$ and $E(y^2)$. If the assembly is " pure " with respect to y, i.e. if all the systems in α have the same value of y, then

$$E(y^2) = [E(y)]^2.$$

If the assembly is not pure the deviation of the values of y from their mean value $\bar{y} = E(y)$ is measured by

$$\begin{aligned} E([y - \bar{y}]^2) &= E(y^2) - 2\bar{y}E(y) + \bar{y}^2 \\ &= E(y^2) - \bar{y}^2. \end{aligned}$$

The square root of this expression,

$$\Delta y = [E(y^2) - \bar{y}^2]^{\frac{1}{2}},$$

is called the "root mean square deviation" or the "uncertainty" in the variable y. If I is the identical operator, and Y is the operator representing y, $[Y - \bar{y}I]^2$ represents $[y - \bar{y}]^2$. Hence

$$(\Delta y)^2 = (\alpha, [Y - \bar{y}I]^2\alpha).$$

As an illustration of the concept of "uncertainty" consider the distribution of kinetic energy among the molecules of a gas. According to Maxwell's distribution law the fraction of the total number of molecules of a gas at absolute temperature T which have kinetic energy W lying in the interval W, W + dW is

$$2\pi(\pi kT)^{-3/2} \exp(-W/kT) \,.\, W^{\frac{1}{2}}dW$$
$$= 4\pi^{-\frac{1}{2}} \exp(-x^2) \,.\, x^2dx,$$

k being Boltzmann's constant and x^2 denoting W/kT. Hence the average value of W is

$$\bar{W} = E(W) = 4\pi^{-\frac{1}{2}} \,.\, kT\int_0^\infty \exp(-x^2) \,.\, x^4dx = \frac{3}{2}kT,$$

and the average value of W^2 is

$$E(W^2) = 4\pi^{-\frac{1}{2}}(kT)^2\int_0^\infty \exp(-x^2)x^6dx = \frac{15}{4}(kT)^2.$$

Therefore the uncertainty in W is

$$\Delta W = [E(W^2) - \bar{W}^2]^{\frac{1}{2}} = (3/2)^{\frac{1}{2}}kT.$$

To return to the main argument, if p and q are two incompatible variables, it will be impossible to separate from a given assembly α one which is "pure" with respect to both p and q. An assembly which is homogeneous in p will exhibit a wide distribution of values of q, and *vice versa*. It may, of course, be possible to remove this disparity in p and q by preparing an assembly which shall exhibit only a small uncertainty in each variable. Hence it is a matter of interest to enquire the relation between the degrees of uncertainty in p and q for any assembly α.

The answer to this question is given in terms of the operator
$$C = i(PQ - QP) = [P, Q]$$
(where $i^2 = -1$). Since
$$(PQ\alpha, \alpha) = (Q\alpha, P\alpha) = (\alpha, QP\alpha)$$
and
$$(iPQ\alpha, \alpha) = -i(PQ\alpha, \alpha) = -(\alpha, iQP\alpha),$$
it follows that
$$(C\alpha, \alpha) = (\alpha, C\alpha),$$
so that C is symmetric. C will here be called the "commutator" of P and Q. Clearly, it is also the commutator of $P + aI$, $Q + bI$, where a and b are ordinary real numbers. The main property of C required here is that
$$\tfrac{1}{4}(\alpha, C\alpha)^2 \leqslant (\alpha, P^2\alpha)(\alpha, Q^2\alpha).$$

To prove this let
$$(P\alpha, Q\alpha) = x + iy,$$
x and y being ordinary real numbers.

Then
$$\begin{aligned}\tfrac{1}{2}(\alpha, C\alpha) &= \tfrac{1}{2}i\{(Q\alpha, P\alpha) - (P\alpha, Q\alpha)\}\\ &= \tfrac{1}{2}i\{(x - iy) - (x + iy)\}\\ &= y \leqslant (x^2 + y^2)^{\frac{1}{2}}.\end{aligned}$$

Hence
$$\begin{aligned}\tfrac{1}{4}(\alpha, C\alpha)^2 &\leqslant (P\alpha, Q\alpha) \,.\, (Q\alpha, P\alpha)\\ &\leqslant (P\alpha, P\alpha) \,.\, (Q\alpha \,.\, Q\alpha),\end{aligned}$$
by Ex. 1, page 11.

Therefore
$$\tfrac{1}{4}(\alpha, C\alpha)^2 \leqslant (\alpha, P^2\alpha)(\alpha, Q^2\alpha).$$

Now $(\Delta p)^2 = (\alpha, [P - \bar{p}I]^2\alpha) = (\alpha, P'^2\alpha)$, say,

and $(\Delta q)^2 = (\alpha, [Q - \bar{q}I]^2\alpha) = (\alpha, Q'^2\alpha)$.

But C is also the commutator of P′ and Q′.

Hence
$$\begin{aligned}\tfrac{1}{4}(\alpha, C\alpha)^2 &\leqslant (\alpha, P'^2\alpha)(\alpha, Q'^2\alpha)\\ &= (\Delta p)^2(\Delta q)^2.\end{aligned}$$

Therefore
$$\Delta p \,.\, \Delta q \geqslant \tfrac{1}{2}(\alpha, C\alpha) = \tfrac{1}{2}E_\alpha(C).$$

This inequality determines the relation between the uncertainties in *any* two variables p and q in terms of the mean value of their commutator C. It shows that homogeneity with respect to one variable, say q, implies an infinite uncertainty in any incompatible variable p.

EXAMPLES :—

(1) If $(\alpha, \alpha)(\beta, \beta) = (\alpha, \beta)(\beta, \alpha)$, prove that there is a non-vanishing complex number c such that $\alpha = c\beta$.

(2) If the uncertainty in y is zero for the assembly α, deduce from the equation $(\alpha, y\alpha)^2 = (\alpha, y^2\alpha)$ that α is a proper vector of y.

(3) Prove that there cannot exist an assembly α such that $\Delta(y) = 0$ for all variables y.

(4) If $\Delta p \,.\, \Delta q = \frac{1}{2}\mathrm{E}(\mathrm{C})$ for an assembly α, prove that $\mathrm{P}'\alpha = ic\mathrm{Q}'\alpha$, and $\Delta p/\Delta q = |\, c \,|$, where c is real and not zero.

(5) Show that if A, B, and C are any three operators, then $[\mathrm{A}, \mathrm{B}] + [\mathrm{B}, \mathrm{A}] = 0$, and

$$[\mathrm{A}, [\mathrm{B}, \mathrm{C}]] + [\mathrm{B}, [\mathrm{C}, \mathrm{A}]] + [\mathrm{C}, [\mathrm{A}, \mathrm{B}]] = 0.$$

(6) Show that

$$[\mathrm{AB}, \mathrm{C}] = \mathrm{A}[\mathrm{B}, \mathrm{C}] + [\mathrm{A}, \mathrm{C}], \mathrm{B},$$

and express $[\mathrm{A}^{-1}, \mathrm{B}]$ in terms of $[\mathrm{A}, \mathrm{B}]$.

The Correspondence Principle.—It follows from the results of the last section that the further detailed development of the theory of "uncertainty" requires explicit expressions for the commutators $\mathrm{C} = i(\mathrm{PQ} - \mathrm{QP})$. These expressions are provided by Heisenberg's exchange relations which will be obtained in the next section.

From the standpoint of the physicist the most satisfactory method of obtaining the exchange relations is by a frank appeal to the Correspondence Principle. This principle is simply a provisional hypothesis adopted to facilitate the functioning of the quantum theory pending deeper investigations into its internal structure. Broadly speaking, the hypothesis is that the characteristics of microphysical systems are expressed by variables of a type similar to those which describe macrophysical systems, i.e. by a set of positional co-ordinates, together with their associated momenta. Thus, for example, we shall assume that the characteristics of the simplest system, a single, structureless particle, are three Cartesian co-ordinates, x, y, z, and the three corresponding components of momentum, p, q, r. This assumption is very simple and plausible, but it definitely transcends our empirical knowledge, and it is independent of the general laws of microphysical measurement formulated in Chapter II.

The operators corresponding to the variables (x, p; y, q; z, r) will be denoted by (X, P; Y, Q; Z, R), writing them in "conjugate pairs," and our problem is to evaluate the commutator of every pair of these operators.

The Principle of Symmetry.—In determining the commutators we shall be guided by the "principle of symmetry," viz. that rotations of the Cartesian frame of reference will leave the commutators unaltered. For example, the transformation

$$x \to x, \quad y \to z, \quad z \to -y,$$
$$p \to p, \quad q \to r, \quad r \to -q,$$

shows that

$$[\mathrm{X}, \mathrm{Y}] = [\mathrm{X}, \mathrm{Z}];$$
$$[\mathrm{P}, \mathrm{Y}] = [\mathrm{P}, \mathrm{Z}],$$
$$[\mathrm{Q}, \mathrm{X}] = [\mathrm{R}, \mathrm{X}].$$

By means of suitable transformations we can now infer that

$$[\mathrm{X}, \mathrm{Y}] = [\mathrm{X}, \mathrm{Z}] = [\mathrm{Y}, \mathrm{Z}] = [\mathrm{Y}, \mathrm{X}].$$

But $\quad [\mathrm{X}, \mathrm{Y}] = -[\mathrm{Y}, \mathrm{X}]$,

whence $\quad [\mathrm{X}, \mathrm{Y}] = 0$,

i.e. any pair of the positional operators commute. Similarly, we can show that

$$[\mathrm{P}, \mathrm{Q}] = 0,$$

i.e. any pair of the momentum operators commute. Again, we can show that

$$[Q, Z] = [P, Z] = [P, Y] = [R, Y],$$

i.e. $$[Q, Z] = [R, Y],$$

and similarly, $$[R, X] = [P, Z],$$
$$[P, Y] = [Q, X].$$

We also note that

$$[P, X] = [Q, Y] = [R, Z].$$

We now employ the transformation

$$x \rightarrow x, \quad y \rightarrow y \cos\theta - z \sin\theta, \quad z \rightarrow y \sin\theta + z \cos\theta,$$

whence

$$\begin{aligned}[Q, Z] &= [Q \cos\theta - R \sin\theta, Y \sin\theta + Z \cos\theta] \\ &= \{[Q, Y] - [R, Z]\} \cos\theta \sin\theta + \\ &\qquad [Q, Z] \cos^2\theta - [R, Y] \sin^2\theta,\end{aligned}$$

i.e. $$[Q, Z] \sin^2\theta = -[R, Y] \sin^2\theta.$$

But we have shown that

$$[Q, Z] = [R, Y].$$

Therefore $$[Q, Z] = 0,$$

and similarly the commutators of each pair of *non-conjugate* operators is zero.

Hence the only commutators which have not been proved to vanish are

$$[P, X] = [Q, Y] = [R, Z] = K, \text{ say.}$$

Now, if A, B, C are any three operators, and A commutes with B and with C, it will also commute with [B, C]. Hence [P, X] commutes with Q, R; Y, Z; and [Q, Y] commutes with R, P; Z, X. Hence the commutator K commutes with all of the six operators X, P; Y, Q; Z, R.

The Exchange Relations.—Three possibilities now arise.

1. K is the nul operator. Then every pair of operators

commute and there is no incompatibility or uncertainty, i.e. microphysics is identical with macrophysics.

2. K is a numerical multiple of the identical operator, i.e. $K = hI$. Clearly h must be a real number, since K is a symmetrical operator. This yields Heisenberg's exchange relations.

3. K is neither the nul operator O nor a numerical multiple of the identical operator I. In this case, since K commutes with every operator, viz. the positional and momentum operators and *every* combination of these, it can still be regarded as an ordinary number K, but the numerical value of K will be indeterminable. Hence the operational character of K will interfere with the determination of the proper values of operators representing prescribed functions of the co-ordinates and momenta, and it will make it impossible to determine average values *absolutely* (i.e. only ratios from which K has been cancelled will be calculable).

We shall adopt the second of these possibilities and recognising that K has the dimensions of action (i.e. the dimensions of energy × time or of angular momentum) we shall be prepared to identify h with a numerical multiple of Planck's constant. The expressions for the commutators may now be collected as

$$\begin{aligned} &[X, Y] = 0, \ldots, \quad [P, Q] = 0, \ldots, \\ &[P, Y] = 0, \ldots, \quad [X, Q], = 0, \ldots, \\ &[P, X] = hI, \ldots \end{aligned}$$

These are Heisenberg's exchange relations.

EXAMPLES :—

(1) Calculate $\Delta p \,.\, \Delta x$ for a pair of conjugate variables.

(2) Show by mathematical induction that

$$[P, X^n] = hnX^{n-1},$$

and $$[P^n, X] = hnP^{n-1},$$

if n is a positive integer.

(3) Show that if $f(x)$ is a polynomial, and $f(x) \to f(X)$, $f'(x) \to f'(X)$, etc., then

$$[P, f(X)] = hf'(X),$$

and $$[f(P), X] = hf'(P).$$

(4) Deduce from Heisenberg's exchange relations that P and X cannot both have a matrix representation in a space with a finite, or even an enumerably infinite number of dimensions. [Form the matrix elements of [P, X], taking as basis the proper vectors of P or X.]

The Wave Operators.—A direct discussion of the matrix representation of a pair of conjugate operators, such as P and X, presents many difficulties, which may, however, be evaded by a preliminary study of the group of unitary operators, $U(a) = \exp(ia\,P/h)$.

It follows from Example 2 of the last section that

$$[U\ a), X] = iaU(a),$$

i.e.

$$U(a\ X = (X + a)U(a).$$

Take as a basis the vectors $\xi_1, \xi_2, \ldots$, where ξ_k is the vector for the state in which $k\delta \leqslant x < (k+1)\delta$. Then X is represented by a diagonal matrix, i.e. $X_{jk} = 0$ unless $j = k$.* The kth diagonal element X_{kk} is clearly $k\delta$. On writing $a = \pm\,\delta$ we find that

$$U_{jk}(\delta)\,.\,k\delta = (j+1)\delta\,.\,U_{jk}(\delta),$$

and

$$U_{jk}(-\delta)\,.\,k\delta = (j-1)\delta\,.\,U_{jk}(-\delta).$$

Hence, $U_{jk}(\delta) = 0$ unless $k = j+1$, and $U_{jk}(-\delta) = 0$ unless $k = j-1$. Now

$$U(\delta)\,.\,U(-\delta) = I.$$

Therefore

$$U_{j,\,j+1}(\delta)\,.\,U_{j+1,\,j}(-\delta) = 1.$$

But $U_{jk}(\delta)$ and $U_{kj}(-\delta)$ are conjugate complex numbers (see p. 17). Therefore the matrix elements of $U(\delta)$, $U_{j,\,j+1}(\delta)$, are complex numbers of modulus unity, of the form $\exp(i\theta_j)$, where θ_j is real.

Now let us construct a set of real numbers $c_1, c_2, \ldots$, such that $c_{j+1} - c_j = \theta_j$, and define a new vector basis $\xi_1', \xi_2', \ldots$, by the equations

$$\xi_k = \exp(ic_k)\,.\,\xi_k'.$$

Then

$$\begin{aligned} U_{j,\,j+1}(\delta) &= (\xi_j, U(\delta)\xi_{j+1}) \\ &= \exp i(c_{j+1} - c_j)\,.\,(\xi_j', U(\delta)\xi'_{j+1}). \end{aligned}$$

* Henceforward it will be convenient to write the j, k-matrix element of X as X_{jk}.

Hence, the non-zero matrix elements of $U(\delta)$ referred to the new basis are all equal to unity. Now the vectors of the new basis represent the same states as the vectors of the old basis. Hence it follows that we can form a representation in which X is represented by the diagonal matrix,

$$\begin{matrix} . & . & . & . & . & \\ . & 1\delta & . & . & . & \\ . & . & 2\delta & . & . & \\ . & . & . & 3\delta & . & \text{etc.} \\ . & . & . & . & 4\delta & \end{matrix}$$

and $U(\delta)$ by the matrix

$$\begin{matrix} . & 1 & . & . & . & \\ . & . & 1 & . & . & \\ . & . & . & 1 & . & \text{etc.} \\ . & . & . & . & 1 & \end{matrix}$$

(the dots representing the zero elements).

Now $U(\delta)\xi_k' = \xi'_{k-1}$ and $U(-\delta)\xi_k' = \xi'_{k+1}$. Hence if $f_1, f_2, \ldots$ are the components of any vector ϕ referred to the new basis, $\xi_1', \xi_2', \ldots$, the components of $U(\delta)\phi$ are $f_2, f_3, \ldots$ (for $f_k = (\xi_k', \phi)$ and $f_{k+1} = (\xi_k', U(\delta)\phi)$). It follows by induction that the kth component of $U(n\delta)\phi$ is f_{k+n}. Now let us fix $n\delta$ equal to a and $k\delta$ equal to x, and proceed to the limit $\delta \to 0$ as on page 38. Since $f_k \to \psi(x)(dx)^{\frac{1}{2}}$, it follows that

$$U(a)\psi(x) = \psi(x+a),$$

i.e.

$$\exp(iaP/h)\psi(x) = \psi(x+a).$$

Now

$$\exp(iaP/h)\psi(x) = \{1 + \sum_{n=1}^{\infty}(iaP/h)^n/n!\}\psi(x),$$

and

$$\psi(x+a) = \{1 + \sum_{n=1}^{\infty}(a^n/n!)(\partial/\partial x)^n\}\psi(x).$$

Therefore

$$P\psi(x) = (h/i)\partial\psi/\partial x,$$

i.e. P is equivalent to the differential operator

$$(h/i)(\partial/\partial x).$$

Also since $X\xi_k' = x_k\xi_k'$, it follows that

$$X\psi(x) = x\psi(x),$$

so that the operator X is equivalent to multiplication by x.

The representation of the action of P and X on wave functions by the "wave operators" $(h/i)(\partial/\partial x)$ and x is due to Schrödinger and forms the basis of his Wave Mechanics, which is a powerful mathematical weapon for calculating the proper values and matrix elements of operators.

EXAMPLES :—

(1) Verify directly that Schrödinger's wave operators satisfy Heisenberg's exchange relation.

(2) If $U(a) = \exp\ (iaP/h)$, $V(b) = \exp\ (ibX/b)$ and η is any vector, prove that the vectors $U(a)V(b)\eta$ and $V(b)U(a)\eta$ represent the same state.

(3) If $\Delta p\ .\ \Delta x$ has its minimum value $\frac{1}{2}h$ for a pair of conjugate variables p and x, prove that the wave function $\psi(x)$ of the corresponding state satisfies the equation

$$\frac{h}{i}\frac{\partial\psi}{\partial x} - \bar{p}\ .\ \psi = ic(x - \bar{x})\psi,$$

where c is real.

Hence deduce that

$$\psi = a \exp \{-\tfrac{1}{2}c(x - \bar{x})^2/h + i\bar{p}x/h\},$$

where

$$|\,a\,| = (c/\pi h)^{\frac{1}{4}}.$$

The Angular Momentum Operators.—The application of quantum theory to problems of atomic and molecular structure is dominated by the theory of the angular momentum operators. If p_1, p_2, p_3 are the components of linear momentum of a structureless particle at the point which Cartesian co-ordinates x_1, x_2, x_3, the components of angular momentum about the origin are

$$m_1 = x_2p_3 - x_3p_2,$$
$$m_2 = x_3p_1 - x_1p_3,$$
$$m_3 = x_1p_2 - x_2p_1.$$

The corresponding operators determined by the rule of page 36 and by the exchange relations are

$$\begin{aligned} M_1 &= \tfrac{1}{2}(X_2P_3 + P_3X_2) - \tfrac{1}{2}(X_3P_2 + P_2X_3) \\ &= X_2P_3 - X_3P_2 = P_3X_2 - P_2X_3, \\ M_2 &= X_3P_1 - X_1P_3, \text{ and } M_3 = X_1P_2 - X_2P_1. \end{aligned}$$

To obtain Schrödinger's representation of these operators as wave operators we introduce the spherical polar co-ordinates r, θ, ϕ defined by the equations

$$\begin{aligned} x_1 &= r \sin\theta \cos\phi, \\ x_2 &= r \sin\theta \sin\phi, \\ x_3 &= r \cos\theta. \end{aligned}$$

Then $$M_3\psi(r, \theta, \phi) = \frac{h}{i}\left\{x_1\frac{\partial\psi}{\partial x_2} - x_2\frac{\partial\psi}{\partial x_1}\right\} = \frac{h}{i}\frac{\partial\psi}{\partial\phi}.$$

The expressions for m_1 and m_2 are given below in Example 1; they are much less important than the expression for m_3, which is especially significant because it shows that ϕ and m_3 are a conjugate pair of variables.

It follows that if

$$W_j(a) = \exp(iaM_j/h),$$

then $$W_3(a) \,.\, \psi(r, \theta, \phi) = \psi(r, \theta, \phi + a).$$

Hence, if the wave function is expressed in Cartesian co-ordinates,

$$\begin{aligned} W_3(a) \,.\, \psi(x, y, z) &\\ &= W(a) \,.\, \psi(r \sin\theta \cos\phi, r \sin\theta \cos\phi, r\cos\theta) \\ &= \psi\,(r \sin\theta \cos\overline{\phi + \alpha}, r\sin\theta \sin\overline{\phi + a}, r\cos\theta) \\ &= \psi(x_1 \cos a - x_2 \sin a, x_2 \cos a + x_1 \sin a, x_j), \end{aligned}$$

i.e. the operator $W_3(a)$ rotates the point (x_1, x_2, x_3) about the z-axis through an angle a.

EXAMPLES :—

(1) Show that

$$M_1 \rightarrow \frac{h}{i}\left\{-\sin\phi\frac{\partial}{\partial\theta} - \cos\phi\cot\theta\frac{\partial}{\partial\phi}\right\},$$

$$M_2 \rightarrow \frac{h}{i}\left\{\cos\phi\frac{\partial}{\partial\theta} - \sin\phi\cot\theta\frac{\partial}{\partial\phi}\right\},$$

and

$$M_1^2 + M_2^2 + M_3^2 \rightarrow -h^2\left\{\frac{1}{\sin\theta}\frac{\partial}{\partial\theta}\left(\sin\theta\frac{\partial}{\partial\theta}\right) + \frac{1}{\sin^2\theta}\frac{\partial^2}{\partial\phi^2}\right\}.$$

(2) Show that

$$X_1 M_1 + X_2M_2 + X_3M_3 = M_1X_1 + M_2X_2 + M_3X_3 = 0,$$

and

$$P_1M_1 + P_2M_2 + P_3M_3 = M_1P_1 + M_2P_2 + M_3P_3 = 0.$$

(3) Show that M_1, M_2 and M_3 each commute with R, (where $R^2 = x_1^2 + x_2^2 + x_3^2$).

The Rotational Exchange Relations.—The "rotational" property of the unitary operator $W_3(a)$ allows us to evaluate, in a simple and direct manner, a number of important commutators, namely, the commutators of the angular momentum operators and the operators representing *any* scalar or the components of *any* vector. The expressions for these commutators, which may be called the "rotational exchange relations," are fundamental in the theory of the selection and intensity rules in atomic spectra developed in Chapter IV.

We first consider the transformation of any scalar $s(x_1, x_2, x_3)$ and the components $v_1(x_1, x_2, x_3)$, $v_2(x_1, x_2, x_3)$, $v_3(x_1, x_2, x_3)$ of any vector when the co-ordinates are subjected to the transformation Chapter IV.

$$x_1' = x_1 \cos a - x_2 \sin a,$$
$$x_2' = x_1 \sin a + x_2 \cos a,$$
$$x_3' = x_3.$$

Clearly,

$$s'(x_1', x_2', x_3') = s(x_1, x_2, x_3),$$
$$v_1'(x_1', x_2' \; x_3') = v_1(x_1, x_2, x_3) \cos a - v_2(x_1, x_2, x_3) \sin a,$$
$$v_2'(x_1', x_2', x_3') = v_1(x_1, x_2, x_3) \sin a + v_2(x_1, x_2, x_3) \cos a,$$
$$v_3'(x_1', x_2', x_3') = v_3(x_1, x_2, x_3).$$

Hence,

$$S' = S.$$
$$V_1' = V_1 \cos a - V_2 \sin a,$$
$$V_2' = V_1 \sin a + V_2 \cos a,$$
$$V_3' = V_3.$$

Secondly, we obtain the transformation of any operator T when the wave functions are subjected to the transformation

$$\psi(x_1', x_2', x_3') = W_3(a)\psi(x_1, x_2, x_3).$$

If T becomes T′, then the wave function $T'\psi(x_1', x_2', x_3')$ must be the transformation of the wave function $T\psi(x_1, x_2, x_3)$, i.e.

$$T'\psi(x_1', x_2', x_3') = W_3(a)T\psi(x_1, x_2, x_3),$$

or $$T'W_3(a)\psi(x_1, x_2, x_3) = W_3(a)T\psi(x_1, x_2, x_3),$$

whence $$T'W_3(a) = W_3(a)T,$$

and $$T' = W_3(a)TW_3(-a).$$

If T′ is expanded in powers of a the leading terms in the expression are

$$\begin{aligned} T' &\sim (1 + iaM_3/h)T(1 - iaM_3/h) \\ &\sim T + (ia/h)(M_3T - TM_3), \end{aligned}$$

i.e. $$[M_3, T] = h \lim_{a\to 0} (T' - T)/a.$$

Finally, we compare the expressions for S', v_1', v_2', v_3' obtained by these two methods. Expanding these operators in powers of a we have

$$\begin{aligned} O = S' - S &= (ia/h)(M_3S - SM_3), \\ -V_2a = V_1' - V_1 &= (ia/h)(M_3V_1 - V_1M_3), \\ V_1a = V_2' - V_2 &= (ia/h)(M_3V_2 - V_2M_3), \\ O = V_3' - V_3 &= (ia/h)(M_3V_3 - V_3M_3). \end{aligned}$$

Therefore,

$$\begin{aligned} M_3S - SM_3 &= O, \\ M_3V_1 - V_1M_3 &= ihV_2, \\ M_3V_2 - V_2M_3 &= -ihV_1, \\ M_3V_3 - V_3M_3 &= O. \end{aligned}$$

Two other groups of four equations each, involving M_1 and M_2 respectively, can be obtained by cyclic permutation of the suffices 1, 2 and 3. It is important to note that these relations are valid for *any* scalar and for *any* vector.

In particular we may take the vector to be the momentum vector itself, whence

$$\begin{aligned} M_2M_3 - M_3M_2 &= ihM_1, \\ M_3M_1 - M_1M_3 &= ihM_2, \\ M_1M_2 - M_2M_1 &= ihM_3. \end{aligned}$$

EXAMPLES :—

(1) Deduce the expressions for the commutators $[M_j, V_k]$ from Heisenberg's exchange relations in the cases (*a*) $V_k = M_k$, (*b*) $V_k = X_k$, (*c*) $V_k = P_k$.

(2) Deduce from the co-ordinate transformation (p. 51) that the operators (i/h) (M_1, M_2, M_3) can be represented by the matrices D_1, D_2, D_3.

$$\left\| \begin{matrix} \cdot & \cdot & \cdot \\ \cdot & \cdot & +1 \\ \cdot & -1 & \cdot \end{matrix} \right\|, \quad \left\| \begin{matrix} \cdot & \cdot & -1 \\ \cdot & \cdot & \cdot \\ +1 & \cdot & \cdot \end{matrix} \right\|, \quad \left\| \begin{matrix} \cdot & +1 & \cdot \\ -1 & \cdot & \cdot \\ \cdot & \cdot & \cdot \end{matrix} \right\|,$$

and hence verify the relations

$$D_2D_3 - D_3D_2 = D_1, \text{ etc.}$$

(3) Show that the matrices

$$S_1 = \left\| \begin{matrix} 0 & 1 \\ 1 & 0 \end{matrix} \right\|, \quad S_2 = \left\| \begin{matrix} 0 & -i \\ i & 0 \end{matrix} \right\|, \quad S_3 = \left\| \begin{matrix} 1 & 0 \\ 0 & -1 \end{matrix} \right\|,$$

satisfy the equations,

$$S_2S_3 = iS_1 = - S_3S_2, \text{ etc.}$$

Hence show that $\frac{1}{2}h(S_1, S_2, S_3)$ satisfy the exchange relations (*supra*) for the momentum vector.

The Matrix Representation of the Angular Momentum Operators, M_1, M_2, M_3.—Schrödinger's representation of the operator M_3 in the form $-ih\partial/\partial\phi$ shows at once that if mh is a proper value of M_3 and if ψ is the wave function for the corresponding proper state, then

$$W_3(a)\psi = \exp(iaM_3/h)\psi = \exp(iam)\psi.$$

But $W_3(2\pi) = I$, whence $\exp(i2\pi m) = 1$, i.e. m is a positive integer or zero, and the proper values of M_3 are

$$mh = (0, \pm 1, \pm 2, \ldots)h.$$

Let us now endeavour to construct a representation of the operators M_1, M_2, M_3 taking as the basis the $(2l + 1)$ proper vectors of M_3 with proper values

$$mh = (0, \pm 1, \pm 2, \ldots, \pm l)h$$

The physical significance of the number l becomes clearer if we consider the operator $M_1{}^2 + M_2{}^2 + M_3{}^2$ which represents the square of the resultant angular momentum. This operator represents a scalar and it therefore commutes with each of the operators M_1, M_2 and M_3. We can therefore choose our basis so that each of the basic vectors is also a proper vector of $M_1{}^2 + M_2{}^2 + M_3{}^2$, in each case with the same proper value, say μ. Then M_3 is represented by a diagonal matrix with diagonal elements $(-l, -l+1, \ldots, l-1, l)h$, and $M_1{}^2 + M_2{}^2 + M_3{}^2$ is represented by μ times the unit matrix.

To determine the value of μ in terms of l we note that the "characteristics" of the matrices representing $M_1{}^2$, $M_2{}^2$, $M_3{}^2$ are each equal to $\frac{1}{3}l(l + 1)(2l + 1)h^2$. For $M_3{}^2$ is represented by a diagonal matrix with diagonal elements

$$(l^2, (l-1)^2, \ldots, (l-1)^2, l^2)h^2,$$

whence

$$\chi(M_3{}^2) = 0 + 2(1^2 + 2^2 + \ldots + l^2)h^2 = \tfrac{1}{3}l(l+1)(2l+1)h^2.$$

Also, since

$$\begin{aligned} W_2(\tfrac{1}{2}\pi)M_3W_2(-\tfrac{1}{2}\pi) &= W_2(+\tfrac{1}{2}\pi)(X_1P_2 - X_2P_1)W_2(-\tfrac{1}{2}\pi) \\ &= (-X_3)P_2 + X_2(P_3) = M_1, \end{aligned}$$

therefore,

$$W_2(\tfrac{1}{2}\pi)M_3{}^2W_2(-\tfrac{1}{2}\pi) = M_1{}^2,$$

and

$$\chi(M_3{}^2) = \chi(M_1{}^2).$$

Similarly, $$\chi(M_3{}^2) = \chi(M_2{}^2).$$

Hence

$$\chi(M_1{}^2 + M_2{}^2 + M_3{}^2) = l(l+1)(2l+1)h^2.$$

But, since $M_1{}^2 + M_2{}^2 + M_3{}^2$ is represented by μ times the unit matrix which has $(2l + 1)$ rows and columns, therefore

$$\chi(M_1{}^2 + M_2{}^2 + M_3{}^2) = \mu(2l + 1).$$

Hence $$\mu = l(l+1)h^2,$$

i.e. the proper values of $M_1{}^2 + M_2{}^2 + M_3{}^2$ are of the form $l(l + 1)h^2$, where l is a positive integer or zero. "l" is called the "serial quantum number" of the proper state.

To determine the matrix elements of M_1 and M_2 it is convenient to introduce the operators

$$M' = M_1 + iM_2 \quad \text{and} \quad M'' = M_1 - iM_2.$$

Then

$$M_3M' - M'M_3 = hM',$$
$$M_3M'' - M''M_3 = -hM'',$$
and $$M'M'' = (M_1{}^2 + M_2{}^2 + M_3{}^2) + hM_3 - M_3{}^2.$$

On replacing the operators occurring in the first and second equations by their corresponding matrices, we find that

$$mM'(m, n) - M'(m, n)n = M'(m, n),$$
$$mM''(m, n) - M''(m, n)n = -M''(m, n).$$

Hence $$M'(m, n) = 0 \quad \text{unless} \quad n = m - 1,$$
and $$M''(m, n) = 0 \quad \text{unless} \quad n = m + 1.$$

We now find from the third operational equation above that

$$\begin{aligned} M'(m, m-1)M''(m-1, m) &= l(l+1)h^2 + mh^2 - m^2h^2 \\ &= [(l+\tfrac{1}{2})^2 - (m-\tfrac{1}{2})^2]h^2 \\ &= (l+m)(l-m+1)h^2. \end{aligned}$$

Now, since M_1 and M_2 are symmetric operators,

$$\begin{aligned} M'(m, n) &= M_1(m, n) + iM_2(m, n) \\ &= M_1^*(n, m) + iM_2^*(n, m) = [M''(n, m)]^*. \end{aligned}$$

Therefore $M'(m, m-1)$ and $M''(m-1, m)$ are conjugate complex numbers. Hence we may write

$$\begin{aligned} M'(m, m-1) &= \epsilon(m)(l+m)^{\frac{1}{2}}(l-m+1)^{\frac{1}{2}}h, \\ M''(m-1, m) &= \epsilon^*(m)(l+m)^{\frac{1}{2}}(l-m+1)^{\frac{1}{2}}h, \end{aligned}$$

where $\epsilon(m)$ is a complex number of modulus unity. Finally, as in the section on Schrödinger's wave operators (p. 47), we can change the basis in such a way that $\epsilon(m)$ becomes equal to unity, the representations of M_3 and $M_1{}^2 + M_2{}^2 + M_3{}^2$ being unchanged. Then $\epsilon^*(m)$ will also be unity and in the new basis

$$M'(m, m-1) = (l+m)^{\frac{1}{2}}(l-m+1)^{\frac{1}{2}}h = M''(m-1, m).$$

Examples :—

(1) Prove directly from the relations

$$M_2M_3 - M_3M_2 = ihM_1, \text{ etc.,}$$

that $M_1{}^2 + M_2{}^2 + M_3{}^2$ commutes with M_3.

(2) Construct explicitly the matrix representation of M_1, M_2, and M_3 in the cases $l = 0$ and 1.

The Reflexion Operator.—The operator R, which is defined by the equation

$$R\psi(x_1, x_2, x_3) = \psi(-x_1, -x_2, -x_3),$$

is called the "reflexion" operator, and is of some importance in the classification of atomic spectra. It is clear that R commutes with the angular momentum operators M_1, M_2, M_3, and that, since $R^2 = I$, its proper values are ± 1. We shall now prove that, in a proper state of M_3 and $M_1{}^2 + M_2{}^2 + M_3{}^2$, in which these operators have the proper values mh and $l(l+1)h^2$, the proper value of R is

$$\begin{aligned} r = (-1)^l &= +1 \text{ if } l \text{ is even,} \\ &= -1 \text{ if } l \text{ is odd.} \end{aligned}$$

This proper value of R is called the "signature" of the proper state.

On adopting Schrödinger's representation of these operators we find that the wave function ψ satisfies the equations

$$\frac{1}{\sin\theta}\frac{\partial}{\partial\theta}\left\{\sin\theta\frac{\partial\psi}{\partial\theta}\right\}+\frac{1}{\sin^2\theta}\frac{\partial^2\psi}{\partial\phi^2}=-l(l+1)\psi.$$

and
$$\frac{\partial\psi}{\partial\phi}=im\psi.$$

Hence ψ involves ϕ only through the factor, exp $im\phi$, and it may therefore be written in the form

$$\psi=e^{im\phi}\,.\,(\sin\theta)^{+m}\,.\,\mathrm{P}_{lm}(\cos\theta).$$

The function P_{lm} (cos θ) satisfies the equation

$$(1-z^2)\frac{d^2\mathrm{P}_{lm}}{dz^2}-2(m+1)z\frac{d\mathrm{P}_{lm}}{dz}$$
$$+[l(l+1)-m(m+1)]\mathrm{P}_{lm}=0,$$

where z has been written for cos θ.

If $m=-l$, this equation is satisfied by the function $\mathrm{P}_l=(1-z^2)^l$, i.e.

$$(1-z^2)\mathrm{P}_l''+2(l-1)z\mathrm{P}_l'+2l\mathrm{P}=0.$$

On differentiating this equation $(l+m)$ times we find that the function

$$\mathrm{F}_{lm}=\frac{d^{l+m}}{dz^{l+m}}\mathrm{P}_l$$

satisfies the equation for P_{lm}. The other solution of this equation becomes infinite as $z\to\pm 1$. Hence the wave function ψ must be a numerical multiple of

$$\mathrm{Y}_{lm}=e^{im\phi}\,.\,(1-z^2)^{\frac{1}{2}m}\,.\,\frac{d^{l+m}}{dz^{l+m}}(1-z^2)^l$$
$$=e^{im\phi}(\sin\theta)^m\mathrm{F}_{lm}.$$

Now F_{lm} is a polynomial in z which contains only even powers if $l+m$ is even, and only odd powers if $l+m$ is odd. Therefore

$$R \,.\, F_{lm} = (-1)^{(l+m)}F_{lm}.$$

Also,

$$e^{im\phi} \,.\, (\sin\theta)^m = (x_1 + ix_2)^m/(x_1{}^2 + x_2{}^2 + x_3{}^2)^{\frac{1}{2}m}.$$

Hence

$$R \,.\, e^{im\phi}\,(\sin\theta)^m = (-1)^m e^{im\phi}\,(\sin\theta)^m.$$

Therefore

$$R \,.\, Y_{lm} = (-1)^{(l+2m)}Y_{lm} = (-1)^l Y_{lm},$$

i.e the proper value of R is $(-1)^l$.

EXAMPLES :—

(1) Show by repeated integration by parts that

$$\int_0^\pi \sin\theta\, d\theta \,.\, \int_0^{2\pi} Y_{lm}^* Y_{l'm'} d\phi = 0$$

unless $l' = l$ and $m' = m$.

(2) Show similarly that

$$\int_0^\pi |Y_{lm}/2^l l\,!\,|^2 \sin\theta\, d\theta = 2(l+m)\,!\,/(l-m)\,!\,(2l+1).$$

The Equations of Motion.—The exchange relations which have been developed in the first half of this chapter express the kinematic properties of microphysical systems. We now have to complete the outline of the quantum theory sketched in this chapter by obtaining analogous expressions for the kinetic properties, i.e. by deducing the quantum equations of motion.

In classical dynamics the " equation of motion " of a variable y is simply an explicit expression for the velocity or time derivative of y, $\dot{y} \equiv dy/dt$, in terms of the instantaneous value of y and the contemporaneous values of the other variables of the system. In quantum theory the "equation of motion" of an operator Y is an explicit expression for an associated operator called by analogy the "velocity of Y."

This concept of the velocity of an operator requires elucidation as, strictly speaking, neither the proper values nor the proper states of a variable suffer any change with lapse of time. Nevertheless, it is convenient to describe a certain operator Z as "the velocity of Y" if the average value of the associated variable z at any instant t is equal to the time rate of change of the average value of y *for any state whatsoever*. If ψ_t is the wave function describing the state at any time t, this definition implies that

$$(\psi_t, \mathrm{Z}\psi_t) = d(\psi_t, \mathrm{Y}\psi_t)/dt.$$

As in classical dynamics we shall write $\mathrm{Z} = \dot{\mathrm{Y}}$.

If the historical development of a system is expressed in the form

$$\psi_t = \mathrm{F}(t)\psi_0,$$

$\dot{\mathrm{Y}}$ can be expressed in terms of Y and F(t). Since we must have

$$(\psi_t, \psi_t) = 1 = (\psi_0, \psi_0),$$

the operator F(t) is unitary (see Ex. 1, p. 19) and is therefore expressible in the form

$$\mathrm{F}(t) = \exp(-it\mathrm{W}/h),$$

where W is a symmetric operator.* Now

$$d\psi_t/dt = (-i/h)\mathrm{W}\psi_t,$$

whence

$$d(\psi_t, \mathrm{Y}\psi_t)/dt = (i/h)(\mathrm{W}\psi_t, \mathrm{Y}\psi_t) + (i/h)(\psi_t, -\mathrm{YW}\psi_t).$$

Therefore $\quad \dot{\mathrm{Y}} = (i/h)(\mathrm{WY} - \mathrm{YW}).$

This expression for $\dot{\mathrm{Y}}$ is here purely formal, as the nature of the operator W is as yet unknown.

* The argument of the exponential has been chosen in anticipation of the subsequent identification of W.

We can now identify the operator W by an appeal to the Correspondence Principle. Clearly the variable w represented by W has the dimensions of energy. We shall show that w is the Hamiltonian function of classical dynamics, i.e. the total energy expressed in terms of the co-ordinates and the momenta.

Consider a particle of mass μ. The kinetic energy is

$$\tfrac{1}{2}\mu(\dot{x}_1^2 + \dot{x}_2^2 + \dot{x}_3^2) = (1/2\mu)(p_1^2 + p_2^2 + p_3^2).$$

Hence if the potential energy is $u(x_1, x_2, x_3)$ the Hamiltonian function is

$$f = (1/2\mu)(p_1^2 + p_2^2 + p_3^2) + u(x_1, x_2, x_3).$$

The " equations of motion " are

$$\begin{aligned} \dot{x}_1 &= p_1/\mu = \partial f/\partial p_1, \ldots, \\ \dot{p}_1 &= -\partial u/\partial x_1 = -\partial f/\partial x_1, \ldots . \end{aligned}$$

Guided by the Correspondence Principle we shall assume that the quantum equations of motion are the operational equivalents of the classical equations, i.e. that the operator representing $\dot{x}_1$ equals the operator representing $\partial f/\partial p_1$, etc.

Now, by the equations of motion obtained above,

$$\dot{x}_1 \to (i/h)(\mathrm{WX}_1 - \mathrm{X}_1\mathrm{W}), \quad \dot{p}_1 \to (i/h)(\mathrm{WP}_1 - \mathrm{P}_1\mathrm{W}).$$

Also, by Heisenberg's exchange relations (p. 46),

$$\partial f/\partial p_1 \to -(i/h)(\mathrm{X}_1\mathrm{F} - \mathrm{FX}_1), \partial f/\partial x_1 \to (i/h)(\mathrm{P}_1\mathrm{F} - \mathrm{FP}_1).$$

Hence (W − F) commutes with X_1, P_1, etc. We shall therefore identify W with F, and can now use the relation

$$\dot{\mathrm{Y}} = (i/h)(\mathrm{WY} - \mathrm{YW}),$$

as an *effective* equation of motion for Y, giving $\dot{\mathrm{Y}}$ in terms of operators representing *known* variables.

EXAMPLES :—

(1) Show that the wave function for a particle with energy w and components of momentum $(p, 0, 0)$ is a numerical multiple of

$$\exp (i/h)(px - wt),$$

i.e. it has frequency $w/2\pi h$ and wave-length $2\pi h/p$. (The form of the wave function in this special case indicates the origin of its name and of the wave theory due to de Broglie. See "Wave Mechanics" in this series by Dr. H. T. Flint.)

(2) If $w = p^2/2\mu + u(x)$, show that

$$\mathrm{P} = \mu\dot{\mathrm{X}} \quad \text{and} \quad \dot{\mathrm{P}} = (i/h)(\mathrm{UP} - \mathrm{PU}).$$

(3) Show that the average value of the energy $\mathrm{E}(w)$ is constant in any state ψ, stationary or non-stationary. Show also that the same is true for any variable y which is compatible with w.

(4) Show that in any proper state of the energy w the average value of the velocity of any variable is zero [i.e. if $\mathrm{W}\psi = w\psi$, $\mathrm{E}\psi(\dot{\mathrm{Y}}) = (\psi, \dot{\mathrm{Y}}\psi) = 0$].

(5) Show that the velocity of YZ is $\dot{\mathrm{Y}}\mathrm{Z} + \mathrm{Y}\dot{\mathrm{Z}}$.

(6) If $w = (1/2\mu)(p_1{}^2 + p_2{}^2 + p_3{}^2) + u(x_1, x_2, x_3)$, where u is a homogeneous function of x_1, x_2, x_3 of the nth degree, prove that in any proper state of w,

$$\mathrm{E}(u) = 2w/(n + 2).$$

$$[d(p_1q_1 + p_2q_2 + p_3q_3)/dt \rightarrow 2\mathrm{W} - (n + 2)\mathrm{U}.]$$

The Harmonic Oscillator.—One of the simplest examples of the determination of the stationary states of a system is provided by the one-dimensional harmonic oscillator. The classical Hamiltonian function for this system is

$$w = p^2/2\mu + \tfrac{1}{2}\mu\omega^2x^2,$$

where μ is its mass and ω the natural pulsatance of the system (i.e. the frequency is $\omega/2\pi$). The corresponding Hamiltonian operator is

$$W = P^2/2\mu + \tfrac{1}{2}\mu\omega^2 X^2,$$

and the equations of motion are therefore

$$\dot{X} = (i/h)(WX - XW) = P/\mu,$$

and

$$\dot{P} = (i/h)(WP - PW) = -\mu\omega^2 X.$$

To solve these equations (i.e. to determine the matrix elements of W, X and P in a stationary state) we note that $\dot{P} + c\dot{X}$ is a numerical multiple of $P + cX$ if the number c is chosen to be $\pm i\mu\omega$. Hence we introduce the operators

$$A = P + i\mu\omega X \quad \text{and} \quad B = P - i\mu\omega X,$$

for which the equations of motion take the simple forms

$$\dot{A} = (i/h)(WA - AW) = i\omega A,$$

and

$$\dot{B} = (i/h)(WB - BW) = -i\omega B.$$

We take as the basis of our representation the set of vectors, $\eta_0, \eta_1, \eta_2, \ldots, \eta_k, \ldots$ which are proper vectors of W, and we denote the corresponding proper values by $w_0, w_1, w_2, \ldots, w_k, \ldots$. These numbers cannot be negative, for

$$2\mu w_k = \sum_n P_{kn}P_{nk} + \mu^2\omega^2 \sum_n X_{kn}X_{nk},$$

and (P_{kn}, P_{nk}), (X_{kn}, X_{nk}) are both pairs of conjugate complex numbers. Hence the set of proper values has a least term, which we shall denote by w_0.

In matrix form the equations of motion are

$$w_j A_{jk} - A_{jk} w_k = h\omega A_{jk},$$

and

$$w_j B_{jk} - B_{jk} w_k = -h\omega B_{jk}.$$

Hence

$$A_{jk} = 0 \quad \text{unless} \quad w_k = w_j - h\omega,$$

$$B_{jk} = 0 \quad \text{unless} \quad w_k = w_j + h\omega,$$

and

$$A_{0k} = 0, \quad B_{j0} = 0.$$

Now $\quad AB = P^2 + \mu^2\omega^2X^2 - i\mu\omega(PX - XP)$
$= 2\mu W - \mu h\omega I,$

i.e. $\quad \sum_n A_{jn}B_{nj} = 2\mu w_j - \mu h\omega,$

and, in particular, for $j = 0$,

$$0 = 2\mu w_0 - \mu h\omega.$$

These results show that the proper values of W form an arithmetic progression whose general term is

$$(n + \tfrac{1}{2})h\omega.$$

Denoting this expression by w_n (as is now permissible), we see that the only non-zero matrix elements of A and B are of the form $A_{n,\,n-1}$ and $B_{n,\,n+1}$. Moreover, it is clear that

$$B_{kj} = P_{kj} - i\mu\omega X_{kj} = P^*_{jk} + (i\mu\omega X_{jk})^* = A^*_{jk},$$

i.e. A_{jk} and B_{kj} are conjugate complex numbers. But

$$\begin{aligned} 2n\mu h\omega &= 2\mu w_n - \mu h\omega \\ &= A_{n,\,n-1}\,B_{n-1,\,n}. \end{aligned}$$

Therefore

$$\begin{aligned} A_{n,\,n-1} &= \epsilon(n)\,.\,(2n\mu h\omega)^{\frac{1}{2}}, \\ B_{n,\,n+1} &= \epsilon^*(n+1)\,.\,[2(n+1)\mu h\omega]^{\frac{1}{2}}, \end{aligned}$$

where $\epsilon(n)$ is a complex number of modulus unity. These results complete the solution of the problem.

EXAMPLES :—

(1) Show that the basis, $\eta_0, \eta_1, \ldots \eta_j, \ldots$, used above, can be chosen so that $\epsilon(n) = 1$.

(2) Verify the result of Example 6, page 61, for the harmonic oscillator, i.e. show that

$$E(\tfrac{1}{2}\mu\omega^2x^2) = \tfrac{1}{2}w_n$$

in the proper state η_n.

(3) Show that the kinetic energy of a particle moving in 3 dimensions is represented by the operator $(1/2\mu)[P_r^{\,2} + (M_1^{\,2} + M_2^{\,2} + M_3^{\,2})/R^2]$, where R represents $r = (x_1^{\,2} + x_2^{\,2} + x_3^{\,2})^{\frac{1}{2}}$, and P_r represents the radial component of momentum. [$2P_r = (X_1/R)P_1 + P_1(X_1/R) + \ldots$.]

(4) Discuss the stationary states of the three dimensional harmonic oscillator, for which

$$W = W_1 + W_2 + W_3,$$

where

$$W_j = P_j^2/2\mu + 2\mu\omega^2 X_j^2,$$

taking as the basis of the representation, (*a*) the proper vectors common to W_1, W_2, W_3, (*b*) the proper vectors common to W, $M_1^2 + M_2^2 + M_3^2$ and M_3.

The Main Properties of the Hamiltonian Operator.— The results of the last section but one are summarised in the definition of the velocity of Y by the formula,

$$(\psi_t, \dot{Y}\psi_t) = d(\psi_t, Y\psi_t)/dt, \qquad \text{(all } \psi_t\text{)},$$

in the equation of motion,

$$\dot{Y} = (i/h)(WY - YW),$$

and in the Schrödinger representation of W as a differential operator,

$$W\psi = (-h/i)\partial\psi/\partial t.$$

The main physical properties of the Hamiltonian operator, W, relate to stationary states and to transitions from one non-stationary state to another.

As regards stationary states, if ψ_t is the wave function for such a state at time t, then ψ_t can differ from ψ_o only by a phase factor of the form exp $[-i\omega(t)/h]$, i.e.

$$\psi_t = \psi_o \,.\, \exp\,[-i\omega(t)/h].$$

Hence, since

$$W = (-h/i)\partial/\partial t,$$

$$W\psi_t = \dot{\omega}(t)\psi_t,$$

and

$$E(w) = \dot{\omega}(t).$$

Now the average value of the energy w is always a constant, as proved in Example 3 above. Hence $\dot{\omega}(t) =$ a constant, w, and

$$\psi_t = \psi_o \,.\, \exp\,(-iwt/h),$$

i.e. ψ_t is a proper wave function of W with proper value w. Therefore any stationary state is a proper state of the energy.

The problem of transitions is as follows: Let the initial state of the system be a proper state, α_j, of a variable a with operator A: what is the probability that the system will be in the proper state, α_k, of the variable a after the lapse of time t? The solution is given by the fact that the actual state of the system after time t is represented by the vector

$$\beta_j = \mathrm{F}(t)\alpha_j = \exp(-it\mathrm{W}/h) \,.\, \alpha_j,$$

whence the required transition probability is

$$|(\beta_j, \alpha_k)|^2 = |(\alpha_k, \mathrm{F}(t)\alpha_j)|^2,$$

i.e. the square of the modulus of the α_j, α_k-matrix element of $\mathrm{F}(t)$.

For sufficiently small intervals of time we may write

$$\mathrm{F}(t) \doteqdot 1 - it\mathrm{W}/h,$$

whence the transition probability is approximately

$$|(\alpha_k, (-it\mathrm{W}/h)\alpha_j)|^2 = |\mathrm{W}_{jk}|^2 \,.\, t^2/h^2,$$

using W_{jk} to denote the j, k-matrix element of W in the basis $\{\alpha_j\}$, i.e. at first the transition probability increases as the square of the time. In particular we note that, if the matrix element W_{jk} vanishes, then the initial transition probability is of order t^4, i.e. it is small in comparison with the transition probabilities for non-vanishing matrix elements of W.

EXAMPLES :—

(1) Show that the probability that a system remains in the same state, α_j, is

$$1 - [(\mathrm{W}^2)_{jj} - (\mathrm{W}_{jj})^2]t^2/h^2$$

for small values of t, and verify that the total probability for all transitions ($\alpha_j \rightarrow \alpha_k$, $k = 1$ 2, . . ., j, . . .) is unity.

(2) If the components of the vector β_j in the basis $\{\alpha_k\}$ are denoted by $b_k = (\alpha_k, \beta_j)$, show that the equations of motion of these components are

$$\dot{b}_k = (-i/h)\sum_n W_{kn} b_n.$$

Using the initial conditions,

$$b_j = 1, \ b_k = 0 \quad \text{if} \quad k \neq j, \ (t = 0),$$

show that

$$b_j = 1 - (iW_{jj}t/h) - \tfrac{1}{2}(W^2)_{jj}(t/h)^2 + \ldots,$$
$$b_k = -(iW_{kj}t/h) + \ldots, \qquad (k \neq j),$$

and hence verify the expressions for the transition probabilities.

Radiative Transitions.—The most important application of the preceding theory is to the transitions which are produced in atomic systems by the absorption or emission of radiation. The complete theory is necessarily lengthy and intricate, but here again an appeal to the Correspondence Principle provides a short (if precarious !) route to the correct result.

According to classical electromagnetic theory a particle of charge e (e.s.u.) moving with an acceleration f radiates energy at the rate $2e^2f^2/3c^3$. If the three co-ordinates of the particle are x_1, x_2, x_3, the average rate of emission of energy will be the average value of

$$(2e^2/3c^3)(\ddot{x}_1{}^2 + \ddot{x}_2{}^2 + \ddot{x}_3{}^2).$$

In Bohr's theory, which occupies an intermediate place between the classical theory and the modern quantum theory, the analogue of this expression for the rate of radiation of energy is an expression for the probability that the particle will leave its initial state "a", and lose one unit of energy in one second.

In the present form of the quantum theory the average value of a variable x in a state "a" is the value of the a, a-matrix element of the corresponding operator X; and the average value of x^2 is similarly

$$E(x^2) = (X^2)_{aa} = \sum_c X_{ac}X_{ca} = \sum_c |X_{ac}|^2.$$

Now, since $\dot{X} = (i/h)(WX - XW)$,

the a, c-matrix element of $\dot{X}$ is

$$(\dot{X}_{ac}) = (i/h)(w_a - w_c)X_{ac},$$

and the a, c-matrix element of $\ddot{X}$ is

$$(\ddot{X})_{ac} = -(1/h)^2(w_a - w_c)^2X_{ac}.$$

Hence $$E(\ddot{x})^2 = \sum_c \omega_{ac}^4 \mid X_{ac} \mid^2,$$

where $$h\omega_{ac} = w_a - w_c.$$

We should therefore expect that the probability of a radiative transition from the state "a," with loss of unit energy in unit time, would be

$$(2e^2/3c^3)\sum_c \omega_{ac}^4\{\mid X_{1,\,ac} \mid^2 + \mid X_{2,\,ac} \mid^2 + \mid X_{3,\,ac} \mid^2\}.$$

Moreover, we should expect that the individual terms in this sum, e.g.

$$(2e^2/3c^3)\omega_{ac}^4\{\mid X_{1,\,ac} \mid^2 + \mid X_{2,\,ac} \mid_2 + \mid X_{3,\,ac} \mid^2\},$$

represent the probability of a radiative transition from the state "a" to individual final states, e.g. the state "c." In the transition $a \rightarrow c$, the energy lost is $w_a - w_c = h\omega_{ac}$. Hence, it appears that the probability of the transition $a \rightarrow c$, with the loss of energy $h\omega_{ac}$ in unit time, is

$$(2e^2\omega_{ac}^3/3c^3h)\{\mid X_{1,\,ac} \mid^2 + \mid X_{2,\,ac} \mid^2 + \mid X_{3,\,ac} \mid^2\}.$$

Otherwise expressed, this is the fraction of the total number of systems in the state "a" which pass over into the state "b" per second, or Einstein's "coefficient of spontaneous emission," $A_{a \rightarrow b}$.

Ignoring the coefficient $(2e^2\omega_{ac}^3/3c^3h)$ (which is actually only one-half of the accepted value), we are led to expect that the probability of the transition $a \rightarrow c$ will depend upon the sum of the three terms,

$$\mid X_{1,\,ac} \mid^2 + \mid X_{2,\,ac} \mid^2 + \mid X_{3,\,ac} \mid^3.$$

We mention this result, not because we attach any value to the method by which it has been derived, but in order to give some physical reality to the calculation of the matrix elements of x_1, x_2, x_3 carried out in the next chapter.

The Identification of *h*.—We noted in the preceding section that a, c-matrix element of $\ddot{\mathrm{X}}$ had the form

$$(\ddot{\mathrm{X}})_{ac} = -\,\omega^2{}_{ac}\,\mathrm{X}_{ac},$$

where
$$\omega_{ac} = (w_a - w_c)/h.$$

This result suggests, by analogy with the classical equations for simple harmonic motion, that during the transition $a \to c$, the variable x is oscillating with pulsatance ω_{ac}, i.e. with frequency $\nu_{ac} = \omega_{ac}/2\pi$. In these circumstances the frequency of the emitted radiation would also be ν_{ac}. Now the energy radiated is

$$w_a - w_c = h\omega_{ac} = (2\pi h)\,\nu_{ac}.$$

But in Bohr's theory of radiation the coefficient of the frequency in this expression is taken to be Planck's constant, i.e. the constant occurring in Planck's formula for the distribution of energy in complete radiation. Hence we are led to identify our constant h with Planck's constant divided by 2π.

EXAMPLES :—

(1) By an appeal to the Correspondence Principle show that if $\mathrm{X}_{1,\,ac} = 0$, $\mathrm{X}_{2,\,ac} = 0$, the character of the radiation from an atom during the transition $a \to c$ is the same as from a particle oscillating along the X_3-axis, i.e. linearly polarised parallel to the x_3-axis (π) when viewed along the x_1- or x_2-axes, but of zero intensity when viewed along the x_3-axis.

(2) Show, similarly, that if $(\mathrm{X}_1 + i\mathrm{X}_2)_{ac} = 0$ or $(\mathrm{X}_1 - i\mathrm{X}_2)_{ac} = 0$, the radiation is circularly polarised when viewed along the x_3-axis (λ) and (ρ), and linearly polarised perpendicular to the X_3-axis (σ) when viewed along the x_1- or x_2-axes.

[The symbols π, σ, λ, ρ, indicating the type of polarisation, denote parallel, perpendicular (senkrecht), left- and right-handed.]

CHAPTER IV

THE SPIN OPERATORS

THE exchange relations and the laws of motion obtained in the last chapter form the basis of the quantum dynamics of a particle. In the systematic application of these general principles to the particular problems of atomic physics it is necessary to adopt a standard method of representing the various scalar and vector functions which characterise atomic systems. Accordingly, we shall discuss in this chapter the standard representation, simplifying the analysis by a systematic use of certain " spin " operators. The results obtained—the proper values and matrix elements of the various operators—are of fundamental importance in atomic theory. The " spin operators " are introduced to obtain a representation which is independent of the particular co-ordinate system employed. In fact, the method of this chapter is the quantum analogue of the use of vectors in classical dynamics.

The Spin Operators.—The method adopted in the previous chapter for the representation of vectors, such as the angular momentum, suffers from the defect that it is dependent upon the co-ordinate system employed. If v_1, v_2, v_3 and v_1', v_2', v_3' are the components of a vector v in two systems of co-ordinates, the two corresponding sets of operators V_1, V_2, V_3 and V_1', V_2', V_3', are connected by the same relations as those which connect the two sets of components. We shall now show that it is possible to represent any vector v by a single operator V, $(v \rightarrow V)$, *the representation being independent* *of the co-ordinate system*, and satisfying the following conditions :—

(1) If $u \to \mathrm{U}$ and $v \to \mathrm{V}$, then $u + v \to \mathrm{U} + \mathrm{V}$ and $cu \to c\mathrm{U}$ (c being an ordinary number) ;

(2) If u and v are compatible variables, then the scalar product

$$u \,.\, v \to \tfrac{1}{2}(\mathrm{UV} + \mathrm{VU}),$$

and the vector product

$$u \times v \to \tfrac{1}{2}i(\mathrm{VU} - \mathrm{UV}).$$

Assuming that such a representation is possible we can determine its nature by the following argument :—

Let s_1, s_2 and s_3 denote the unit vectors with components (1, 0, 0), (0, 1, 0) and (0, 0, 1) and let S_1, S_2, S_3 denote the corresponding operators. Then the vector l with numerical components (l_1, l_2, l_3) is represented by the operators $l_1\mathrm{S}_1 + l_2\mathrm{S}_2 + l_3\mathrm{S}_3$. Hence, by the second condition,

$$(l_1\mathrm{S}_1 + l_2\mathrm{S}_2 + l_3\mathrm{S}_3)^2 = l_1^{\,2} + l_2^{\,2} + l_3^{\,2},$$

i.e.
$$\mathrm{S}_1^{\,2} = \mathrm{S}_2^{\,2} = \mathrm{S}_3^{\,2} = \mathrm{I},$$

and
$$\mathrm{S}_2\mathrm{S}_3 + \mathrm{S}_3\mathrm{S}_2 = 0, \ . \ . \ ., \text{ etc.}$$

Also S_1 is the vector product of S_2 by S_3. Hence

$$\tfrac{1}{2}i(\mathrm{S}_3\mathrm{S}_2 - \mathrm{S}_2\mathrm{S}_3) = \mathrm{S}_1, \ . \ . \ ., \text{ etc.}$$

Therefore,
$$\mathrm{S}_2\mathrm{S}_3 = i\mathrm{S}_1 = -\,\mathrm{S}_3\mathrm{S}_2.$$

Similarly,
$$\mathrm{S}_3\mathrm{S}_1 = i\mathrm{S}_2 = -\,\mathrm{S}_1\mathrm{S}_3,$$

and
$$\mathrm{S}_1\mathrm{S}_2 = i\mathrm{S}_3 = -\,\mathrm{S}_2\mathrm{S}_1.$$

The three operators, S_1, S_2, S_3 are the " spin operators " introduced by Pauli and Dirac.

We shall now show that the operator representing any vector v is

$$\mathrm{V} = \mathrm{V}_1\mathrm{S}_1 + \mathrm{V}_2\mathrm{S}_2 + \mathrm{V}_3\mathrm{S}_3.$$

In order that V may be a " symmetric " operator it is necessary and sufficient that S_1 should commute with

V_1, S_2 with V_2, and S_3 with V_3 (Ex. 5, p. 12). Assuming the possibility of constructing spin operators to fulfil these conditions, we note that the scalar product $s_1.v$ will be represented by $\frac{1}{2}(S_1V + VS_1) = V_1$, which represents $(v_1, 0, 0)$; and that the vector product $s_2 \times v$ will be represented by $\frac{1}{2}i(S_1V - VS_1) = V_3S_2 - V_2S_3$, which represents $(0, v_3, -v_2)$. Hence V must represent the vector with components (v_1, v_2, v_3). We shall, later, verify the possibility assumed above by actually calculating the matrix elements of S_1, S_2, S_3 in a representation in which these operators commute with X_1, X_2, X_3 and P_1, P_2, P_3.

The invariance of the representation of v by V follows from the fact that the two sets of operators, V_1, V_2, V_3 and S_1, S_2, S_3, both transform like the co-ordinates x_1, x_2, x_3 in any rotation of the co-ordinate system. Hence the " scalar product " of these two sets of operators is invariant, i.e.

$$V' = V_1'S_1' + V_2'S_2' + V_3'S_3' = V_1S_1 + V_2S_2 + V_3S_3 = V.$$

Hence the operator representing V is unchanged by the rotation of the co-ordinates.

The Spin Variable.—The spin operators, unlike the " dynamical operators," do not represent any physical variable; nevertheless, in the mathematical theory they must be treated on the same basis as the dynamical operators in order to obtain a matrix representation of the operator V corresponding to any vector v. Thus we must regard the operator S_3 as representing a fictitious variable w, whose proper values are the proper values of S_3, i.e. $+1$ and -1. This variable w is called the " spin variable." It is merely a mathematical fiction introduced to simplify the analysis.

To adopt this fiction consistently we must now consider a complete observation to be not merely an observation of the positional co-ordinates x_1, x_2 and x_3, but also of the spin variable w. Hence the wave functions must be regarded as functions $\psi(x_1, x_2, x_3, w)$ of the four variables x_1, x_2, x_3, w; or alternately, since

w takes only two values ± 1, we can regard a wave function as a sort of vector * with two components

$$\psi_1(x_1, x_2, x_3) = \psi(x_1, x_2, x_3, 1), \text{ and } \psi_2(x_1, x_2, x_3) = \psi(x_1, x_2, x_3, -1).$$

If we make use of the representation of the spin operators suggested by Example 3, page 53, we find that, on writing

$$\eta_1 = \psi(x_1, x_2, x_3, +1), \quad \eta_2 = \psi(x_1, x_2, x_3, -1),$$

then
$$S_1\eta_1 = \eta_2, \qquad S_1\eta_2 = \eta_1,$$
$$S_2\eta_1 = i\eta_2, \qquad S_2\eta_2 = -i\eta_1,$$
$$S_3\eta_1 = \eta_1, \qquad S_3\eta_2 = -\eta_2.$$

In view of these relations it is clearly unnecessary to introduce other spin variables corresponding to S_1 and S_2 as w does to S_3. The privileged position given to S_3 does introduce a certain lack of symmetry into the theory, but this very lack of symmetry is especially useful in the application of the theory to physical problems in which the occurrence of a magnetic or electrical field introduces a similar directional emphasis.

In expressing the scalar product of two vectors α and β in terms of their wave functions $\psi_\alpha(x_1, x_2, x_3, w)$ and $\psi_\beta(x_1, x_2, x_3, w)$, the summation or integration must now be extended over the domains of all four variables, x_1, x_2, x_3, w. Hence

$$(\beta, \alpha) = \int\{\psi_\beta^*(x_1, x_2, x_3, 1)\psi_\alpha(x_1, x_2, x_3, 1) + \psi_\beta^*(x_1, x_2, x_3, -1)\psi_\alpha(x_1, x_2, x_3, -1)\}\, dx_1, dx_2, dx_3.$$

EXAMPLES :—

(1) If $S' = S_1 + iS_2$, $S'' = S_1 - iS_2$, evaluate $S'\eta_j$ and $S''\eta_j$. $(j = 1, 2.)$

(2) If $S_3\eta_1 = \eta_1$, $S_3\eta_2 = -\eta_2$, find the most general expressions for $S_1\eta_j$, $S_2\eta_j$, consistent with the relations $S_2S_3 = iS$, etc., and with the assumption that the spin operators are "symmetric."

$$[S_1 = e^{ia}\,\eta_2, \quad S_1\eta_2 = e^{-ia}\,\eta_1, \text{ etc.}]$$

* A "semi-vector" (Landau), or a "spinor" (van der Waerden).

(3) If S_3 and S_3' are the spin operators for two axes inclined at an angle a, show that the proper vectors of S_3' are

$$\eta_1' = \eta_1 \cos \tfrac{1}{2}a + \eta_2\, i \sin \tfrac{1}{2}a,$$

and

$$\eta_2' = -\eta_1 i \sin \tfrac{1}{2}a + \eta_2 \cos \tfrac{1}{2}a.$$

Hence show that the probability of a transition from a proper state of S_3 to a proper state of S_3' is $\sin^2 \frac{1}{2}a$ or $\cos^2 \frac{1}{2}a$, according as the proper value does or does not change.

(4) If $T(a) = \cos \frac{1}{2}a + iS_3 \sin \frac{1}{2}a$, show that

$$S_j'T(a) = T(a)S_j, \quad (j = 1, 2, 3),$$

and prove that $\{T(a)\}$ is a group of unitary operators.

The Rotational Exchange Relations.—The introduction of the spin variable necessitates a modification of the rotational exchange relations of page 51, for a rotation of the axes of reference transforms the spin operator S_3 and hence the spin variable w in addition to the positional operators and co-ordinates X_1, X_2, X_3 and x_1, x_2, x_3. The simplest way to obtain the requisite modification is to note that if x_1, x_2, x_3 suffer the transformation,

$$x_1' = x_1 \cos a - x_2 \sin a,$$
$$x_2' = x_1 \sin a + x_2 \cos a, \quad x_3' = x_3',$$

then S_1, S_2, S_3 suffer the same transformation,

$$S_1' = S_1 \cos a - S_2 \sin a,$$
$$S_2' = S_1 \sin a + S_2 \cos a, \quad S_3' = S_3.$$

These equations may be written as

$$S_j' = T(a)S_jT(-a), \quad (j = 1, 2, 3)$$

whence $T(a) = \cos \frac{1}{2}a + iS_3 \sin \frac{1}{2}a = \exp [\frac{1}{2}iaS_3]$.

Now the unitary operator $T(a)$ transforms only the spin operators, while the unitary operator $W_3(a)$ of page 50 transforms only the "dynamical" operators. Hence both types of operators will be transformed by the unitary operator $\exp [ia(M_3 + \frac{1}{2}hS_3)/h]$. Therefore, in the new transformation theory, which takes into account spin operators and spin variables, the operators M_1, M_2 and M_3 of the old theory will be replaced by N_1, N_2 and N_3 where

$$N_j = M_j + \tfrac{1}{2}hS_j, \quad (j = 1, 2, 3).$$

The rotational exchange relations analogous to those of page 52 are

$$\begin{aligned} N_3V_1 - V_1N_3 &= ihN_2, \\ N_3V_2 - V_2N_3 &= -\,ihN_1, \\ N_3V_3 - V_3N_3 &= 0. \end{aligned}$$

These relations are valid for the operators representing the components of any vector V_1, V_2, V_3, and also for the spin operators S_1, S_2, S_3—as may be verified directly by substituting S for V. If, however, we employ the single operator $V = V_1S_1 + V_2S_2 + V_3S_3$ to represent the vector v, then, since this representation is invariant,

$$\exp(iaN_3/h)\,.\,V.\;\; \exp(-\,iaN_3/h) = V' = V,$$

whence it follows that V commutes with N_3. Similarly, V commutes with N_1 and N_2. This simplification of the rotational exchange relations illustrates the advantages of employing the spin operators.

EXAMPLES :—

(1) Evaluate the wave functions

$$\exp(iaN_j/h)\psi(x_1, x_2, x_3, w)$$

for $j = 1, 2, 3$, and hence verify the result of Example 3, page 73.

(2) If the operators B_1, B_2, B_3 and C_1, C_2, C_3 all commute with the spin operators, prove that

$$BC = B_1C_1 + B_2C_2 + B_3C_3 + i\{S_1(B_2C_3 - B_3C_2) + \ldots\}$$

(3) Show that

$$\tfrac{1}{2}i(PX - XP) = N.$$

(4) Prove directly from the definition of N_1, N_2, N_3 that $N_2N_3 - N_3N_2 = ihN_1$, and that $N_3N = NN_3$.

The Spin Momentum.—It follows at once from the rotational exchange relations that the Hamiltonian operator W, being a scalar, must commute either with the operators N_1, N_2, N_3 or with the operators M_1, M_2, M_3, accordingly as W does or does not involve the spin operators. The importance of this result is that it is quite independent of the *structure* of the Hamiltonian operator, and therefore that is much more fundamental than any consequence deduced from the Correspondence Principle (which narrowly restricts the form of W, as we have seen on p. 60).

In fact, the Correspondence Principle is only a temporary expedient which must sooner or later be replaced by a more profound study of the nature of microphysical systems. Hence we shall not hesitate in this chapter to consider the possibility of Hamiltonian operators which involve the spin operators as well as the dynamical operators. The structure of these operators will be discussed briefly at the end of this chapter. Meanwhile, recognising provisionally that W may involve S_1, S_2 and S_3, we note that it is not the operators M_j, but the operators N_j, which will *necessarily* commute with W.

It follows that the angular momentum variables m_1, m_2, m_3 will be incompatible with the energy w, so that it will be impossible for a particle to have a definite energy and definite angular momentum in any one prescribed direction. Indeed, in a stationary state, with a definite, constant value of w, the average values of m_1, m_2, m_3 will vary with the time. In fact, it is the operators N_j which have all the properties which we should expect of the operators M_j. This surprising result suggests the following physical interpretation of the N_j's, viz. that the variables m_1, m_2, m_3 form only a part of the total angular momentum, the other part consisting of the variables $\frac{1}{2}hs_j$ represented by the operators $\frac{1}{2}hS_j$, and the total angular momentum having components represented by the N_j's.

This interpretation becomes more plausible if the first part of the total angular momentum m_1, m_2, m_3 is regarded as arising from the *translational* motion of the particle, and the other part from the *rotational* motion of the particle, which must now be regarded not as a structureless point, but as a body of small but finite size. The first part of the angular momentum can then be described as the " mechanical " momentum, and the second part as the " spin " momentum. There is no need to insist upon this interpretation, although it is sometimes useful in the imaginative construction of a dynamical model of microphysical systems.

The important conclusion to be drawn from this discussion relates to the complete description of the stationary states of a particle.

W commutes not only with N_1, N_2 and N_3 but also with N, which itself commutes with the three N_j's. Hence W, N and N_3, say, form a set of commuting operators, and the stationary states of a particle can be completely described as the proper states of these three operators. The corresponding proper vectors form the basis of the standard representation of the states of a particle.

The structure, and, therefore, the proper values of W will clearly depend upon the external influences acting upon the particle, which will vary from case to case. Hence, in a general treatment, we can discuss only the representation based upon the proper vectors common to N and N_3.

The Matrix Representation of N_1, N_2, N_3.—The matrix representation of the operators N_1, N_2 and N_3 can be determined by methods similar to those employed in Chapter III (p. 53) for the operators M_1, M_2 and M_3.

The proper values of M_3 are of the form

$$mh = (0, \pm 1, \pm 2, \ . \ . \ . \ )h.$$

Hence, since $N_3 = M_3 + \frac{1}{2}hS_3$, $S_3{}^2 = I$, and M_3 commutes with S_3, it follows that the proper values of N_3 are of the form

$$uh = (\pm \tfrac{1}{2}, \pm \tfrac{3}{2}, \pm \tfrac{5}{2}, \ . \ . \ .)h.$$

We then construct a representation by means of matrices of n rows and columns, taking as the basis the proper vectors of N_3 with proper values

$$uh = (\pm \tfrac{1}{2}, \pm \tfrac{3}{2}, \ldots, \pm j)h.$$

As before we find that $N_1{}^2 + N_2{}^2 + N_3{}^2$ commutes with N_3 and that the vectors of the basis can therefore be chosen to be proper vectors of $N_1{}^2 + N_2{}^2 + N_3{}^2$, in each case with the same proper value, which is found to be $j(j + 1)h^2$. Hence the proper values of $N_1{}^2 + N_2{}^2 + N_3{}^2$ are of the form $j(j + 1)h^2$, where j is half an odd positive integer.

Similarly, we easily find that the non-vanishing matrix elements of $N' = N_1 + iN_2$ and $N'' = N_1 - N_2 i$ are of the form $N'(u, u - 1)$ and $N''(u - 1, u)$, and that the basis can be chosen so that

$$N'(u, u - 1) = (j + u)^{\frac{1}{2}}(j - u + 1)^{\frac{1}{2}}h = N''(u - 1, u).$$

For the purposes of classifying atomic spectra it is necessary to adopt a representation which will include the angular momentum operators and the spin operators. But, since S_3 commutes with N_3 and $N_1{}^2 + N_2{}^2 + N_3{}^2$, a basis adequate for this purpose must contain a set of vectors which are simultaneously proper vectors of N_3, $N_1{}^2 + N_2{}^2 + N_3{}^2$ and S_3. Such bases were employed in the older forms of the quantum theory, but in more recent work the calculations are considerably simplified by taking as a basis the proper vectors of

$$N_3 \text{ and } M = M_1S_1 + M_2S_2 + M_3S_3 = N - \tfrac{3}{2}hI.$$

EXAMPLES :—

(1) Prove directly from the rotational exchange relations $M_2M_3 - M_3M_2 = ihM_1$, etc., that the operators S_3, M_3, N_3, $M_1{}^2 + M_2{}^2 + M_3{}^2$, $N_1{}^2 + N_2{}^2 + N_3{}^2$ all commute with one another.

(2) If η is a proper vector of these operators with proper values $2s$, mh, uh, $l(l + 1)h^2$, $j(j + 1)h^2$, for S_3, M_3, N_3, etc., show that

$$\begin{aligned}
j &= l \pm \tfrac{1}{2} \text{ if } l \neq 0 \\
&= \tfrac{1}{2} \text{ if } l = 0\,; \\
u &= m + s, \\
u_{\max} &= l + \tfrac{1}{2},\ u_{\min} = -l - \tfrac{1}{2}, \\
s &= +\tfrac{1}{2} \text{ if } u = u_{\max} \\
&= j - l \text{ if } u_{\min} < u < u_{\max} \\
&= -\tfrac{1}{2} \text{ if } u = u_{\min}.
\end{aligned}$$

The Standard Representation in Terms of N_3 and $K = N - \frac{1}{2}hI$.—It is rather more convenient to replace M by the operator

$$K = M + hI = N - \tfrac{1}{2}hI.$$

Starting from the relation

$$M^2 = M_1{}^2 + M_2{}^2 + M_3{}^2 - hM,$$

we find that

$$M_1{}^2 + M_2{}^2 + M_3{}^2 = K^2 - hK,$$

and

$$\begin{aligned}
N_1{}^2 + N_2{}^2 + N_3{}^2 &= (M_1 + \tfrac{1}{2}hS_1)^2 + \ldots \\
&= M_1{}^2 + M_2{}^2 + M_3{}^2 + hM + \tfrac{3}{4}h^2 I \\
&= K^2 - \tfrac{1}{4}h^2 I.
\end{aligned}$$

We have already proved (p. 74) that N_3 commutes with M and hence with K. It now appears that $M_1{}^2 + M_2{}^2 + M_3{}^2$ and $N_1{}^2 + N_2{}^2 + N_3{}^2$ are expressible as polynomials in K, and therefore that they commute with K and N_3 and with one another.

Now let $l(l+1)h^2$, $j(j+1)h^2$, kh and uh be the proper values of $M_1{}^2 + M_2{}^2 + M_3{}^2$, $N_1{}^2 + N_2{}^2 + N_3{}^2$, K and N_3 for a common proper vector $\eta\ (k, u)$. Then

$$l(l+1) = k(k-1) \quad \text{and} \quad j(j+1) = k^2 - \tfrac{1}{4}.$$

But l and j are never negative. Hence

$$l = |k - \tfrac{1}{2}| - \tfrac{1}{2},\ j = |k| - \tfrac{1}{2},$$

and

$$k = j(j+1) - l(l+1) + \tfrac{1}{4}.$$

The numbers l, j and k, which specify a proper state, are called the "serial," "inner," and "auxiliary" quantum numbers of the state.

Since the possible values of l are 0, 1, 2, . . . , and the possible values of j are $\frac{1}{2}, \frac{3}{2}, \frac{5}{2}, \ldots$, it follows that the possible values of k are $\pm 1, \pm 2, \pm 3, \ldots$, *the value 0 being excluded.* Hence the proper values of K are of the form

$$kh = (\pm 1, \pm 2, \pm 3, \ldots)h.$$

When the value of k is fixed the possible values of u are $-|k| + \frac{1}{2}, -|k| + \frac{3}{2}, \ldots, |k| - \frac{3}{2}, |k| - \frac{1}{2}$, ($2|k|$ in number). We shall take as the standard basis the set of vectors, $\eta(k, u)$, ($-|k| + \frac{1}{2} \leqslant u \leqslant |k| - \frac{1}{2}$, and k being unrestricted), and shall verify that all the operators $M_1, M_2, M_3, S_1, S_2, S_3$ can be represented by matrices in terms of this basis. The matrix elements of any operator T will be specified by means of two pairs of suffixes (k, u) and (k', u'), according to the notation

$$T(k, u\,;\; k', u') = (\eta(k, u), T\eta(k', u')).$$

The commuting operators N_3, K, $M_1^2 + M_2^2 + M_3^2$ and $N_1^2 + N_2^2 + N_3^2$ are all represented by diagonal matrices. No other angular momentum or spin operator commutes with both N_3 and K, so that we have a maximum number of independent, commuting operators.

The representation of N_1 and N_2 can be determined as in the previous section. If $N' = N_1 + iN_2$ and $N'' = N_1 - iN_2$, then N' and N'' both commute with K, so that the only non-vanishing matrix elements are of the form $N'(k, u\,;\; k, u - 1)$ and $N''(k, u - 1\,;\; k, u)$, with the common value

$$[(j + \tfrac{1}{2})^2 - (u - \tfrac{1}{2})^2]^{\frac{1}{2}}h = (k + u - \tfrac{1}{2})^{\frac{1}{2}}(k - u + \tfrac{1}{2})^{\frac{1}{2}}h.$$

The Matrix Representation of the Spin Operators S_1, S_2, S_3.—As in the case of the operators M_1, M_2, M_3 and N_1, N_2, N_3 it will be convenient to introduce the operators

$$S' = S_1 + iS_2 \quad \text{and} \quad S'' = S_1 - iS_2.$$

Then, since S_1 and S_2 are symmetric, the matrix elements

$$S'(k, u\,;\; k', u') \quad \text{and} \quad S''(k', u'\,;\; k, u)$$

are conjugate complex numbers.

Dealing first with S_3, we note that S_3 commutes with N_3. Hence $S_3(k, u; k', u') = 0$ unless $u' = u$. Also S_3 satisfies the equation

$$S_3K + KS_3 = 2M_3 + 2hS_3 = 2N_3 + hS_3.$$

On replacing the operators by matrix elements we find that

$$S_3(k, u; k', u) . k' + k . S_3(k, u; k', u) = 2N_3(k, u; k', u)/h + S_3(k, u; k', u).$$

Hence

$$(2k - 1)S_3(k, u; k, u) = 2u,$$

and

$$(k' + k - 1)S_3(k, u; k', u) = 0, \text{ if } k' \neq k.$$

Therefore the only non-vanishing matrix elements of S_3 are of the form

$$S_3(k, u; k, u) \quad \text{and} \quad S_3(k, u; 1 - k, u).$$

Moreover, the diagonal elements are given by

$$S_3(k, u; k, u) = \frac{u}{k - \frac{1}{2}}.$$

The moduli of the non-diagonal elements can be determined from the equation

$$S_3^2 = I,$$

i.e.

$$S_3^2(k, u; k, u) + S_3(k, u; 1 - k, u)S_3(1 - k, u; k, u) = 1,$$

whence

$$|S_3(k, u; 1 - k, u)| = \left\{1 - \frac{u^2}{(k - \frac{1}{2})^2}\right\}^{\frac{1}{2}} = (k + u - \tfrac{1}{2})^{\frac{1}{2}}(k - u - \tfrac{1}{2})^{\frac{1}{2}}/|k - \tfrac{1}{2}|,$$

(since S_3 is a symmetric operator).

Turning now to the operator S' we form the equations

$$N_3S' - S'N_3 = hS',$$

and

$$S'K + KS' = 2M' + 2hS' = 2N' + hS',$$

from which it follows that

$$S'(k, u\,;\; k', u') = 0 \text{ unless } u' = u - 1 \text{ and } k' = k \text{ or } 1 - k.$$

Moreover, the second equation also shows that

$$\begin{aligned} S'(k, u\,;\; k, u - 1) &= N'(k, u\,;\; k, u - 1)/(k - \tfrac{1}{2})h \\ &= (k + u - \tfrac{1}{2})^{\frac{1}{2}}(k - u + \tfrac{1}{2})^{\frac{1}{2}}/(k - \tfrac{1}{2}). \end{aligned}$$

It follows similarly that

$$S''(k, u\,;\; k', u') = 0 \text{ unless } u' = u - 1 \text{ and } k' = k \text{ or } 1 - k,$$

and

$$S''(k, u - 1\,;\; k, u) = S'(k, u\,;\; k, u - 1).$$

The surviving matrix elements can be deduced from the relation

$$S'S'' = 2I + 2S_3,$$

i.e.

$$\begin{aligned} &S'(k, u\,;\; k, u - 1)S''(k, u - 1\,;\; k, u) \\ &\quad + S'(k, u\,;\; 1 - k, u - 1)S''(1 - k, u - 1\,;\; k, u) \\ &\quad = 2 + \frac{2u}{k - \frac{1}{2}}, \end{aligned}$$

whence

$$\begin{aligned} |\,S'(k, u\,;\; 1 - k, u - 1)\,| &= |\,S''(1 - k, u - 1\,;\; k, u)\,| \\ &= (k + u - \tfrac{1}{2})^{\frac{1}{2}}(k + u - \tfrac{3}{2})^{\frac{1}{2}}/|k - \tfrac{1}{2}|. \end{aligned}$$

The Selection Rules for X_1, X_2, X_3.—The selection rules for any operator are simply the formulæ which give explicitly the non-vanishing matrix elements of that operator in some agreed representation. By means of the rotational exchange relations of page 74 we can easily determine the selection rules for the operators X_1, X_2, X_3 in the standard representation. The importance of these rules is that, as noted on page 67, they determine the possible radiative transitions of a particle, and the character of the radiation emitted.

If r denotes, as usual $(x_1{}^2 + x_2{}^2 + x_3{}^2)^{\frac{1}{2}}$, and R denotes the corresponding operator, it is clear that R commutes with N_3 and K and hence that the basis of the standard representation is inadequate to include R, and therefore X_1, X_2, X_3. We shall therefore consider the operators Y_1, Y_2, Y_3 which represent x_1/r, x_2/r, x_3/r, and which can be represented in the standard way.

Let $Y = Y_1S_1 + Y_2S_2 + Y_3S_3$. Then Y commutes with N_3 (p. 74). Hence

$$Y(k, u\,;\ k', u') = 0 \text{ unless } u' = u.$$

This is the " selection rule " for " u."

Now

$$\begin{aligned} MY + YM &= (M_1Y_1 + \ldots) + (M_2Y_3 - M_3Y_2)S_2S_3 + \ldots \\ &\quad + (Y_2M_3 - Y_3M_2)S_2S_3 + \ldots + (Y_1M_1 + \ldots) \\ &= 0 + (M_2Y_3 - Y_3M_2)iS_1 + \ldots + \\ &\quad + (Y_2M_3 - M_3Y_2)iS_1 + \ldots + 0 \\ &= -2h(Y_1S_1 + Y_2S_2 + Y_3S_3) = -2hY. \end{aligned}$$

Therefore $$KY + YK = 0,$$

i.e. $$Y(k, u\,;\ k', u) = 0 \text{ unless } k' = -k.$$

This is the selection rule for " k."

The selection rules for Y_1, Y_2, Y_3 can now be determined from the equations

$$Y_j = \tfrac{1}{2}(S_jY + YS_j).$$

For example, by the selection rules for Y,

$$\begin{aligned} 2Y_3(k, u\,;\ k', u') &= S_3(k, u\,;\ -k', u')Y(-k', u'\,;\ k', u') \\ &\quad + Y(k, u\,;\ -k, u)S_3(-k, u\,;\ k', u'). \end{aligned}$$

Hence, by the selection rules for S_3,

$$\begin{aligned} Y_3(k, u\,;\ k', u') = 0 \text{ unless } \quad & u' = u, \\ \text{and} \quad & -k' = k \text{ or } 1 - k, \\ \text{or} \quad & k' = -k \text{ or } 1 + k. \end{aligned}$$

Therefore the selection rules for Y_3 are that

$$Y_3(k, u\,;\; k', u') = 0 \text{ unless } u' = u, \quad \text{and} \quad k' = -k \text{ or } k \pm 1.$$

It is easily verified that the same selection rule in k holds for $Y' = Y_1 + iY_2$ and for $Y'' = Y_1 - iY_2$. The selection rules in u for these operators are the same as for S' and S'' respectively, i.e. $u' = u \mp 1$.

In terms of j and l the selection rule in k becomes

$$j' = j, \qquad l' = l \pm 1,$$

or

$$j' = j + 1, \; l' = l + 1,$$

or

$$j' = j - 1, \; l' = l - 1.$$

These are the well-known rules for the inner and serial quantum numbers.

EXAMPLES :—

(1) Establish the following selection rules for an operator T representing *any* scalar :

$T(k, u\,;\; k', u') = 0$ unless $u' = u$ and $k' = \pm k$.

[T commutes with K^2 and N_3.]

(2) If

$$F_1 = KX_3 - X_3K,$$
$$F_2 = KF_1 - F_1K$$
and $$F_3 = KF_2 + F_2K,$$

prove that

$$F_1 = -ihS_1X_2 + ihS_2X_1,$$
$$F_2 = h^2X_3 + h(N_3X + XN_3),$$
$$(X = X_1S_1 + X_2S_2 + X_3S_3),$$
and $$F_3 = h^2(KX_3 + XK_3).$$

Hence show that

$$K^3X_3 - K^2X_3K - KX_3K^2 + X_3K^3 = h^2(KX_3 + X_3K),$$

and deduce that $X_3(k, u\,;\; k', u') = 0$ unless

$$(k + k')(k^2 - 2kk' + k'^2 - 1) = 0, \text{ etc.}$$

(3) Show that the " signature " (i.e. the proper value of the reflexion operator R of p. 56) of any proper state of K is $(-1)^l$. Show also that

the selection rule for the signature remains unaltered by the introduction of the spin operators.

(4) If $G = M_1^2 + M_2^2 + M_3^2$, prove directly from the rotational exchange relations that

$$G^2V_j - 2GV_jG + V_jG^2 = 2h^2(GV_j + V_jG),$$

for each component V_j of any vector and hence deduce the selection rule in l, i.e.

$$[(l + \tfrac{1}{2})^2 - (l' - \tfrac{1}{2})^2][(l + \tfrac{1}{2})^2 - (l' + \tfrac{3}{2})^2] = 0,$$

or $\quad l' = l \pm 1, - l \quad$ or $\quad - l - 2.$

Show also that the non-negative characters of l and l' exclude the fourth possibility, and permit the third possibility only if $l = l' = 0$. Why is this possibility also excluded ?

The Matrix Representation of Y_1, Y_2, Y_3.—The non-vanishing matrix elements of Y are of the form $Y(k, u\,;\; - k, u)$, and we can show that these elements are of the form exp $i\theta_k$, where θ_k is real, and depends only on k. For Y commutes with N′, i.e.

$$Y(k, u\,; - k, u)N'(- k, u\,;\; - k, u - 1)$$
$$= N'(k, u\,;\; k, u - 1)Y(k, u - 1\,; -k, u - 1).$$

Now

$$N'(k, u\,;\; k, u - 1)$$
$$= \{k^2 - (u - \tfrac{1}{2})^2\}^{\frac{1}{2}}h = N'(- k, u\,; - k, u - 1),$$

(see p. 79). Hence

$$Y(k, u\,; - k, u)$$
$$= Y(k, u - 1\,;\; -k, u - 1)$$
$$= Y(k, u - 2\,;\; - k, u - 2), \text{ etc.,}$$

i.e. $Y(k, u\,;\; - k, u)$ is independent of u. Also $Y^2 = I$, whence

$$Y(k, u\,;\; - k, u)Y(- k, u\,;\; k, u) = 1.$$

Since Y is symmetric the factors on the left are conjugate complex numbers, whence $Y(k, u\,;\; - k, u)$ is of the form

$\exp i\theta_k$, where θ_k is an *odd* function of k, i.e. $\theta_{-k} = -\theta_k$. We shall write $\exp i\theta_k = \epsilon(k)$ for brevity.

The absolute values of the matrix elements of Y_3, Y' and Y'' can now be obtained from the relations

$$Y_3 = \tfrac{1}{2}(S_3 Y + Y S_3), \text{ etc.,}$$

and the known absolute values of the matrix elements of S_3, S' and S''.

Thus

$$\begin{aligned} Y_3(k, u\,;\ -k, u) &= \tfrac{1}{2}S_3(k, u\,;\ k, u)Y(k, u\,;\ -k, u) \\ &\quad + \tfrac{1}{2}Y(k, u\,;\ -k, u)S_3(-k, u\,;\ -k, u) \\ &= \tfrac{1}{2}\left\{\frac{u}{k-\frac{1}{2}} + \frac{u}{-k-\frac{1}{2}}\right\}\epsilon(k) \\ &= \frac{\frac{1}{2}u\epsilon(k)}{k^2-\frac{1}{4}}. \end{aligned}$$

The complete set of matrix elements are tabulated below.

$$\underline{k' = -k.}$$

$$Y'(k, u\,;\ -k, u-1) = \tfrac{1}{2}\epsilon(k)(k+u-\tfrac{1}{2})^{\frac{1}{2}}(k-u+\tfrac{1}{2})^{\frac{1}{2}}/(k^2-\tfrac{1}{4}),$$

$$Y''(k, u\,;\ -k, u+1) = \tfrac{1}{2}\epsilon(k)(k+u+\tfrac{1}{2})^{\frac{1}{2}}(k-u-\tfrac{1}{2})^{\frac{1}{2}}/(k^2-\tfrac{1}{4})$$

$$Y_3(k, u\,;\ -k, u) = \tfrac{1}{2}\epsilon(k)u/(k^2-\tfrac{1}{4}).$$

$$\underline{k' = k+1.}$$

$$|Y'(k, u\,;\ k+1, u-1)| = \tfrac{1}{2}(k-u+\tfrac{1}{2})^{\frac{1}{2}}(k-u+\tfrac{3}{2})^{\frac{1}{2}}/|k+\tfrac{1}{2}|,$$

$$|Y''(k, u\,;\ k+1, u+1)| = \tfrac{1}{2}(k+u+\tfrac{3}{2})^{\frac{1}{2}}(k+u+\tfrac{1}{2})^{\frac{1}{2}}/|k+\tfrac{1}{2}|,$$

$$|Y_3(k, u\,;\ k+1, u)| = \tfrac{1}{2}(k-u+\tfrac{1}{2})^{\frac{1}{2}}(k+u+\tfrac{1}{2})^{\frac{1}{2}}/|k+\tfrac{1}{2}|.$$

$$\underline{k' = k-1.}$$

$$|Y'(k, u\,;\ k-1, u-1)| = \tfrac{1}{2}(k+u-\tfrac{1}{2})^{\frac{1}{2}}(k+u-\tfrac{3}{2})^{\frac{1}{2}}/|k-\tfrac{1}{2}|,$$

$$|Y''(k, u\,;\ k-1, u+1)| = \tfrac{1}{2}(k-u-\tfrac{3}{2})^{\frac{1}{2}}(k-u-\tfrac{1}{2})^{\frac{1}{2}}/|k-\tfrac{1}{2}|,$$

$$|Y_3(k, u\,;\ k-1, u)| = \tfrac{1}{2}(k+u-\tfrac{1}{2})^{\frac{1}{2}}(k-u-\tfrac{1}{2})^{\frac{1}{2}}/|k-\tfrac{1}{2}|.$$

EXAMPLES :—

[These examples are exercises on the representation of the operators, V_1, V_2, V_3, representing the components of *any* " dynamical " vector, in the basis formed by the unit vectors, $\alpha(l, m, s)$, such that

$$(M_1^2 + M_2^2 + M_3^2)\alpha(l, m, s) = l(l+1)h^2 \,.\, \alpha(l, m, s),$$
$$M_3\alpha(l, m, s) = mh \,.\, \alpha(l, m, s),$$
$$\tfrac{1}{2}S_3\alpha(l, m, s) = sh \,.\, \alpha(l, m, s).]$$

(1) Show that the selection rules are

$$s' = s,\ l' = l \pm 1$$

for all three components, and

$$m' = m - 1 \text{ for } V' = V_1 + iV_2,$$
$$m' = m + 1 \text{ for } V'' = V_1 - iV_2,$$
$$m' = m \text{ for } V_3.$$

(2) Show that

$$M'V' = V'M',\ M''V'' = V''M'',$$
$$M''V' - V'M'' = -2hV_3,$$
$$M'V'' - V''M' = 2hV_3.$$

Hence deduce that

$$V'(l, m;\ l-1, m-1) = +C(l)\,.\,(l+m)^{\frac{1}{2}}(l+m-1)^{\frac{1}{2}},$$
$$V''(l, m-1;\ l-1, m) = -C(l)\,.\,(l-m)^{\frac{1}{2}}(l-m+1)^{\frac{1}{2}},$$
$$V_3(l, m;\ l-1, m) = -C(l)\,.\,(l-m)^{\frac{1}{2}}(l+m)^{\frac{1}{2}};$$
$$V'(l-1, m;\ l, m-1) = +C^*(l)\,.\,(l-m+1)^{\frac{1}{2}}(l-m)^{\frac{1}{2}},$$
$$V''(l-1, m-1;\ l, m) = -C^*(l)\,.\,(l+m)^{\frac{1}{2}}(l+m-1)^{\frac{1}{2}},$$
$$V_3(l-1, m;\ l, m) = -C^*(l)\,.\,(l-m)^{\frac{1}{2}}(l+m)^{\frac{1}{2}};$$

where $C(l)$ is a function of l only. The matrix elements of M', M'' are given on p. 56. Since S_3 commutes with V_1, V_2, V_3, the quantum number " s " need not be quoted.)

(3) Show (*a*) that the $(l, m;\ l, m)$ matrix element of

$$V_1^2 + V_2^2 + V_3^2 = \tfrac{1}{2}(V'V'' + V''V') + V_3^2$$

is $\quad l(2l-1)|C(l)|^2 + (l+1)(2l+3)|C(l+1)|^2;$

and (*b*) that the $(l, m; l, m)$ matrix element of $V'V'' - V''V'$ is

$$2m(2l-1)|C(l)|^2 - 2m(2l+3)|C(l+1)|^2.$$

(4) The Bohr model of a "hydrogen-like" atom consists of an electron of mass μ and charge $-e$, moving under the Coulomb attraction of a fixed nucleus of charge Ze. Hence the Hamiltonian function for the electron is

$$w = (1/2\mu)(p_1^2 + p_2^2 + p_3^2) - Ze^2/r,$$

where $r^2 = x_1^2 + x_2^2 + x_3^2$. The "eccentricity" vector, whose components are

$$a_1 = (x_1/r) + (m_2p_3 - m_3p_2)/(\mu Ze^2), \ldots,$$

is constant in magnitude and direction. Its magnitude is the eccentricity ϵ of the orbit described by the electron, and its direction is along the major axis. The total energy is given by

$$\mu Z^2e^4(a_1^2 + a_2^2 + a_3^2 - 1) = 2(m_1^2 + m_2^2 + m_3^2)w.$$

Construct the analogue of this argument in quantum dynamics, showing that
(*a*) the operators A_1, A_2, A_3, which represent a_1, a_2, a_3, each commute with the Hamiltonian operator, W;

(*b*) $\mu Z^2e^4(A_1^2 + A_2^2 + A_3^2 - I) = (M_1^2 + M_2^2 + M_3^2 + h^2I)W;$

(*c*) $A_1A_2 - A_2A_1 = (2h/i\mu Ze^2)\, M_3W$, etc.

(5) Show that in the $2(n^2)$-dimensional representation whose basis consists of the vectors, $\alpha(l, m, s)$, such that

$$l = 0, 1, 2, \ldots, n-1,$$
$$m = -l, -l+1, \ldots, l-1, l,$$

W is represented by a multiple w_n, say, of the unit matrix I; and A′, A″, A_3 are represented by the matrices of Example 2 with

$$|C(l)|^2 = \frac{n^2 - l^2}{4l^2 - 1}\left\{-\frac{2h^2 w_n}{\mu Z^2 e^4}\right\}.$$

[Use Examples 3(*b*) and 4(*c*).]
Show also, by Example 4(*b*) that the value of w_n is $-\mu Z^2 e^4/2h^2 n^2$, i.e.

$$w_n = -Z^2 C/n^2,$$

where $C/2\pi h$ is the Rydberg constant.

(It is remarkable that all the rigorous methods of obtaining this formula for the energy levels of the Bohr hydrogen atom require equally lengthy and intricate argument.)

The Relativistic Spin Operators.—The preceding account of the spin operators is incomplete because it deals only with the three spatial components of a vector, and yields a representation which is invariant only with respect to ordinary spatial rotations. A complete theory should be applicable to relativistic 4-vectors (such as the electric charge and current 4-vector or the energy and momentum 4-vector) with three spatial and one temporal component, and should yield a representation invariant with respect both to rotations and to Lorentz transformations. We proceed to give a brief account of this "quantum tensor analysis" necessitated by the demands of the special theory of relativity.

Let E_1, E_2, E_3, E_4 be the operators which represent the 4-vectors with components (1, 0, 0, 0), (0, 1, 0, 0), (0, 0, 1, 0), (0, 0, 0, 1)—(the temporal component is written last). Let V_1, V_2, V_3, V_4 be the four commuting operators, which represent the four components (v_1, v_2, v_3, v_4) of a 4-vector v and let

$$V = V_1E_1 + V_2E_2 + V_3E_3 + iV_4E_4$$

be the operator representing v. Then V^2 must represent the square of the magnitude of v, or the number

$$v^2 = -v_1{}^2 - v_2{}^2 - v_3{}^2 + v_4{}^2,$$

the units of time and distance being chosen to make the invariant velocity equal to unity. Hence we must have

$$E_j{}^2 = -I, \quad E_jE_k + E_kE_j = 0, \ (k \neq j), \ . \ . \ .$$

It follows from these equations that the E_j's cannot be symmetric operators: we may, however, take them to be "skew." If we define E_5 by the equation,

$$E_5 = iE_1E_2E_3E_4,$$

we find that the preceding equations still hold when j and k can take the values 1, 2, 3, 4, 5. All possible products of the E_j's, formed with any number of factors, will now be of the form cE_{jk} $(c = \pm 1, \pm i)$, where

$$E_{jk} = E_jE_k, \quad (j, k = 0, 1, 2, 3, 4, 5),$$

and $E_0 = iI.$

We now find that if u and v are compatible variables then the scalar product, $\boldsymbol{u} \cdot \boldsymbol{v}$ is represented by the operator $\frac{1}{2}(UV + VU)$; while the operator $\frac{1}{2}i(UV - VU)$ is equal to $\sum_{j \neq k}(U_jV_k - V_jU_k)E_{jk}$. This operator evidently represents the 6-vector, whose components are

$$u_2v_3 - u_3v_2 = a_1, \quad u_3v_1 - u_1v_3 = a_2, \quad u_1v_2 - u_2v_1 = a_3,$$

and

$$u_1v_4 - u_4v_1 = b_1, \quad u_2v_4 - u_4v_2 = b_2, \quad u_3v_4 - u_4v_3 = b_3.$$

When we restrict ourselves to ordinary rotations and exclude Lorentz transformations, the two sets of components (a_1, a_2, a_3) and (b_1, b_2, b_3) both behave like the components of two vectors a and b, which are represented by the operators

$$A = A_1S_1 + A_2S_2 + A_3S_3,$$

and $B = B_1T_1 + B_2T_2 + B_3T_3,$

where $S_1 = iE_{23}, S_2 = iE_{31}, S_3 = iE_{12},$

and $T_1 = iE_{14}, T_2 = iE_{24}, T_3 = iE_{34}.$

We now see that

$$S_1^2 = I, \quad S_2S_3 = iS_1 = -S_3S_2, \text{ etc.,}$$
$$T_1^2 = I, \quad T_2T_3 = iS_1 = -T_3T_2, \text{ etc.,}$$

so that the S_j's have all the properties of the non-relativistic spin operators introduced at the beginning of this chapter (p. 70). Moreover, since the components of the angular momentum really form a quasi-vector like a, i.e. since they are the three spatial components of a 6-vector, the representation of the angular momentum by an operator of the form

$$M = M_1S_1 + M_2S_2 + M_3S_3$$

is entirely justified.

EXAMPLES :—

(1) Show that the sixteen operators,

$$E_0, \; E_j, \; E_{jk}, \quad (j, k = 1, 2, 3, 4, 5, j \neq k),$$

are all skew and linearly independent.

(2) If E_1, E_2, E_3, E_4 and E_1', E_2', E_3', E_4' are two similar sets of spin operators, prove that

$$PE'_{jk} = E_{jk}P \quad \text{and} \quad PE_{jk} = E'_{jk}P,$$

where $$P = \tfrac{1}{4}i\Sigma E_{jk}E'_{jk}, \qquad \text{(16 terms)}$$

$$P^{-1} = -\tfrac{1}{4}i\Sigma E'_{jk}E_{jk},$$

and

$$P^2 = -I.$$

(3) If $$R_1 = iE_{45}, \quad R_2 = iE_5, \quad R_3 = iE_4,$$

show that

$$R_1^2 = I, \quad R_2R_3 = iR_1, \text{ etc.}$$

and that R_j commutes with S_k for all values of j and k

(4) Show that the transformation operator P of Example 2 can be factorised in the form

$$P = iP_1P_2$$

where $$P_1 = \tfrac{1}{2}(I + S_1S_1' + S_2S_2' + S_3S_3'),$$
$$P_2 = \tfrac{1}{2}(I + R_1R_1' + R_2R_2' + R_3R_3'),$$

and P_1 commutes with P_2.

The Relativistic Wave Function.—In order to obtain a matrix representation of the operator

$$V = V_1E_1 + V_2E_2 + V_3E_3 + iV_4E_4,$$

we must treat the spin operators E_j on the same footing as the dynamical operators, and we must therefore introduce certain "spin variables" just as in our previous investigation (p. 71). The examples at the end of the preceding section show that the complete set of E_{jk}'s contain two commuting sets of operators R_1, R_2, R_3 and S_1, S_2, S_3, which both behave like Pauli's spin operators. Hence, it will be convenient to take as the basis for the representation of the E_{jk}'s the proper vectors common to R_3 and S_3. The proper values of these two operators are ± 1, and since

$$\chi(R_3) = \chi(R_2R_3R_2) = \chi(-R_3)$$

and

$$\chi(S_3) = \chi(S_2S_3S_2) = \chi(-S_3),$$

it follows that, for each operator, the sum of its proper values is zero.

Let f and g denote the spin variables represented by R_3 and S_3, then the relativistic wave function will have the form

$$\psi(x_1, x_2, x_3, x_4, f, g),$$

when regarded as a function of the positional co-ordinates and the spin co-ordinates. (Here x_4 denotes the time.) It may also be regarded as a pseudo-vector with four components

$$\begin{aligned}\psi_1(x_1, x_2, x_3, x_4) &= \psi(x_1, x_2, x_3, x_4, 1, 1),\\ \psi_2(x_1, x_2, x_3, x_4) &= \psi(x_1, x_2, x_3, x_4, 1, -1),\\ \psi_3(x_1, x_2, x_3, x_4) &= \psi(x_1, x_2, x_3, x_4, -1, 1),\\ \psi_4(x_1, x_2, x_3, x_4) &= \psi(x_1, x_2, x_3, x_4, -1, -1),\end{aligned}$$

since f and g can take only the values ± 1.

For fixed values of the positional co-ordinates, R_3 and S_3 will be represented by the matrices

$$R_3 = \begin{matrix} 1 & . & . & . \\ . & 1 & . & . \\ . & . & -1 & . \\ . & . & . & -1 \end{matrix} \quad \text{and} \quad S_3 = \begin{matrix} 1 & . & . & . \\ . & -1 & . & . \\ . & . & 1 & . \\ . & . & . & -1 \end{matrix}$$

and it will easily be verified that the simplest representation of R_1, R_2, S_1, S_2 is provided by the matrices

$$R_1 = \begin{matrix} \cdot & \cdot & 1 & \cdot \\ \cdot & \cdot & \cdot & 1 \\ 1 & \cdot & \cdot & \cdot \\ \cdot & 1 & \cdot & \cdot \end{matrix}\,, \quad R_2 = \begin{matrix} \cdot & \cdot & -i & \cdot \\ \cdot & \cdot & \cdot & -i \\ i & \cdot & \cdot & \cdot \\ \cdot & i & \cdot & \cdot \end{matrix}$$

$$S_1 = \begin{matrix} \cdot & 1 & \cdot & \cdot \\ 1 & \cdot & \cdot & \cdot \\ \cdot & \cdot & \cdot & 1 \\ \cdot & \cdot & 1 & \cdot \end{matrix}\,, \quad S_2 = \begin{matrix} \cdot & -i & \cdot & \cdot \\ i & \cdot & \cdot & \cdot \\ \cdot & \cdot & \cdot & -i \\ \cdot & \cdot & i & \cdot \end{matrix}$$

The representation of the remaining spin operators can now be deduced from the equations

$$iE_1 = R_1S_1, \quad iE_2 = R_1S_2, \quad iE_3 = R_1S_3, \quad iE_4 = R_3, \text{ etc.}$$

It follows from the nature of this representation that the spin operators commute with the dynamical operators.

EXAMPLES :—

(1) If $\quad X = X_1S_1 + X_2S_2 + X_3S_3$

and $\quad P = P_1S_1 + P_2S_2 + P_3S_3,$

prove that

$$XP = RP_r + iK,$$

where

$$2P_r = \sum_j [(X_j/R)P_j + P_j(X_j/R)],$$

$$= (2/R)(X_1P_1 + X_2P_2 + X_3P_3) - 2ih/R,$$

and $K = M_1S_1 + M_2S_2 + M_3S_3 + hI.$

(2) Prove that R_3K commutes with P. [See Ex. 3, p. 90.]

The Relativistic Hamiltonian Operator.—There is only one point in the preceding account of the principles of the non-relativistic quantum theory which now requires essential modification—namely, the treatment of the Hamiltonian operator. In the non-relativistic theory this operator, representing the energy, was a scalar, and therefore it commuted with the operators N_1, N_2, N_3 and N. But in the relativistic theory the energy w is simply part of the temporal component p_4 of the energy-momentum 4-vector (p_1, p_2, p_3, p_4), and the preceding argument is therefore invalid. We can, however, determine the form of W by an appeal to the Correspondence Principle.

In the classical form of the special theory of relativity the energy-momentum 4-vector is equal in magnitude to the proper mass m_0 of the particle considered, i.e.

$$p_4{}^2 - p_1{}^2 - p_2{}^2 - p_3{}^2 = m_0{}^2.$$

Hence we shall assume that the operator

$$\Pi \equiv \mathrm{P_1E_1 + P_2E_2 + P_3E_3} + i\mathrm{P_4E_4}$$

has proper values $\pm m_0$. For a state with proper value μ we find that

$$\begin{aligned} \mathrm{P_4} &= - i\mathrm{E_4(P_1E_1 + P_2E_2 + P_3E_3)} + i\mathrm{E_4}\mu \\ &= - \mathrm{E_4R_1(P_1S_1 + P_2S_2 + P_3S_3)} + i\mathrm{E_4}\mu \\ &= \mathrm{R_2 \,.\, P + R_3} \,.\, \mu. \end{aligned}$$

Now the total energy w is the sum of the kinetic energy p_4 and the potential energy u. Hence

$$\begin{aligned} \mathrm{W} &= \mathrm{P_4 + U} \\ &= \mathrm{U + R_2P + R_3}\mu. \end{aligned}$$

This is the form of the relativistic Hamiltonian operator, when all the (linear) momentum is mechanical. In a magnetic field part of the momentum arises from the action of the field and a corresponding allowance must be made.

EXAMPLES :—

[The following sequence of exercises apply to the relativistic Hamiltonian operator for hydrogen-like atoms in which $\mathrm{U} = -\mathrm{Z}e^2/\mathrm{R}$. On choosing the unit of time so that the critical velocity is c, the Hamiltonian operator becomes

$$\mathrm{W} = -\mathrm{Z}e^2/\mathrm{R} + \mathrm{R_2P}c + \mathrm{R_3}\mu c^2.$$

The "fine structure constant" e^2/hc is denoted by α.]

(1) Show that the operators W, $\mathrm{R_3K}$ and $\mathrm{G = K} + i(\mathrm{Z}e^2\mathrm{X}/c\mathrm{R})\mathrm{R_2}$ all commute with one another.

(2) If η is a proper vector of W, R_3K and G with proper values w, kh and gh, show that

$$c(X/R)P_r\eta =$$
$$\{-iR_1\mu c^2 + R_2(w + Ze^2/R) - iR_3khc(X/R^2)\}\eta,$$
and $$c^2P_r{}^2\eta =$$
$$\{-\mu^2c^4 + w^2 + 2wZe^2/R - g(g-1)h^2c^2/R^2\}\eta.$$

[P_r commutes with X/R, $P_rR^{-1} - R^{-1}P_r = ihR^{-2}$, and $P = (X/R)(P_r + iK/R)$.]

(3) Show that in the non-relativistic theory of hydrogen-like atoms the analogous relation is of the form

$$P_r{}^2\eta = \{w_0 + a/R - \lambda(\lambda+1)h^2/R^2\}\eta,$$

where $a = Ze^2/R$ and λ is the serial quantum number.

[$P_1{}^2 + P_2{}^2 + P_3{}^2 = P_r{}^2 + (M_1{}^2 + M_2{}^2 + M_3{}^2)/R^2$.]

The proper values of w_0 have been shown (p. 88) to be

$$w_0 = -a^2/4h^2(\lambda + n_0)^2$$
where $$n_0 = 1, 2, 3, \ldots$$

Show from a comparison of the two expressions for $P_r{}^2$, here and in Example 2, that the proper values of w in the relativistic theory are given by

$$\left\{\frac{\mu c^2}{w}\right\}^2 = 1 + \frac{Z^2\alpha^2}{(n_0 + |g - \frac{1}{2}| - \frac{1}{2})^2}.$$

This is Sommerfeld's formula.

(4) Show that $g = k(1 - Z^2\alpha^2/k^2)^{\frac{1}{2}}$, and obtain the following approximate formula for the proper values of w:

$$m_0c^2 - w = \frac{CZ^2}{n^2} + \frac{C\alpha^2Z^4}{n^3}\left\{\frac{1}{j+\frac{1}{2}} - \frac{3}{4n}\right\} + \ldots,$$

where

$$C = m_0e^4/2h^2,\ n = n_0 + l,\ l + \tfrac{1}{2} = |k - \tfrac{1}{2}|,\ j = |k| - \tfrac{1}{2}.$$

CHAPTER V

COMPOSITE SYSTEMS

In Chapters III and IV we have developed the general principles of the quantum dynamics of a single particle. We must now pass on to study the dynamics of systems composed of several particles. The two main problems are the relation of such a system, considered as a whole, to the particles, or sub-systems, of which it is composed ; and the interaction of the sub-systems among themselves, with its effect on the historical development of the system as a whole.

Any serious attempt to solve these problems lies outside the scope of this book. The theory expounded in this chapter is mainly preparatory to the study of the major problems, and is principally concerned with the minor problem of the proper description of the states of composite systems in terms of the maximum number of compatible variables. This investigation is in some sense parallel to the rigid kinematics of classical theory, with its theory of the description of rigid bodies in terms of centres and moments of inertia.

The Exchange Relations.—The first step in the problem of the dependence or independence of sub-systems is the construction of the exchange relations which express the compatibility or incompatibility of variables belonging to *different* sub-systems. There is little difficulty in showing by the principle of symmetry (p. 44) that the operators representing variables belonging to different sub-systems necessarily commute with one another.

Consider first the operators X_1^A, X_2^A, X_3^A and X_1^B, X_2^B, X_3^B representing the Cartesian positional coordinates of two particles A and B. By rotating the co-ordinate system about the point (X_1^A, X_2^A, X_3^A), we can show that

$$(X_1^B - X_1^A, X_1^A) = (X_1^A - X_1^B, X_1^A),$$
$$(X_1^B - X_1^A, X_2^A) = (X_1^A - X_1^B, X_2^B),$$

etc., whence X_1^B commutes with X_1^A, X_2^A, X_3^A. Similarly, we can show that X_1^A, X_2^A, X_3^A all commute with X_2^B, X_3^B.

Secondly, turning to the components of momentum, P_1^A, P_2^A, P_3^A and P_1^B, P_2^B, P_3^B, the method of argument applied on page 44 to the commutators (Q, Z) and (R, Y), can also be applied to $(P_2^A, X_3^B - X_3^A)$ and $(P_3^A, X_2^B - X_2^A)$, thus showing that P_2^A commutes with X_3^B and P_3^A with X_2^B. By similar arguments it follows that P_1^B, P_2^P, P_3^B all commute with X_1^A, X_2^A, X_3^A. Hence we conclude that *any* dynamical variable of one particle commutes with *any* dynamical variable of another particle.

Finally, we must consider the two sets of spin operators S_1^A, S_2^A, S_3^A and S_1^B, S_2^B, S_3^B. If V_1^A, V_2^A, V_3^A and V_1^B, V_2^B, V_3^B are the operators representing the components of two vectors belonging to the two particles, the vectors themselves will be represented by

$$V^A = V_1^A S_1^A + V_2^A S_2^A + V_3^A S_3^A$$

and

$$V^B = V_1^B S_1^B + V_2^B S_2^B + V_3^B S_3^B.$$

It will obviously be convenient to ensure that V^A commutes with V^B, and we shall therefore choose the spin operators so that *any* spin operator of the particle A commutes with *any* spin operator (or dynamical operator) of the particle B, and vice versa.

In this chapter we shall use the notation

$$V_1 = V_1^A + V_1^B + \dots, \quad V_2 = V_2^A + V_2^B + \dots,$$

etc., for the operators representing the sum of the corresponding components of vectors belonging to different particles of a system. In this notation the rotational exchange relations for a composite system are

$$N_3 S^A - S^A N_3 = 0,$$

(for a scalar operator S^A of the particle A),

$$N_3V_1^A - V_1^AN_3 = ihV_2^A,$$
$$N_3V_2^A - V_2^AN_3 = -ihV_1^A,$$
$$N_3V_3^A - V_3^AN_3 = 0, \text{ etc.},$$

where $N_3 = M_1^A + M_2^A + \ldots + \frac{1}{2}h(S_1^A + S_1^B + \ldots)$.

EXAMPLES :—

(1) Prove that in a composite system

$$N_3S - SN_3 = 0, \quad (S = S^A + S^B + \ldots)$$
$$N_3V_1 - V_1N_3 = ihV_2,$$
$$N_3V_2 - V_2N_3 = -ihV_1,$$
$$N_3V_3 - V_3N_3 = 0, \text{ etc.}$$

(2) Prove directly that in a composite system

$$M_2M_3 - M_3M_2 = ihM_1, \text{ etc.},$$

and $$N_2N_3 - N_3N_2 = ihN_1, \text{ etc.}$$

(3) Show that the proper values of N_3 are of the form

$$(0, \pm 1, \pm 2, \ldots)h,$$

or $$(\pm \tfrac{1}{2}, \pm \tfrac{3}{2}, \pm \tfrac{5}{2}, \ldots)h,$$

according as the number of particles is even or odd.

(4) If

$$S^{AB} = \tfrac{1}{2}(1 + S_1^AS_1^B + S_2^AS_2^B + S_3^AS_3^B),$$

show that $S^{AB}S_j^A = S_j^BS^{AB}, \quad (j = 1, 2, 3)$,

and $(S^{AB})^2 = I$.

(5) Show that the spin operators for two particles can be represented by the matrices

$$S_1^A \rightarrow \begin{matrix} . & 1 & . & . \\ 1 & . & . & . \\ . & . & . & 1 \\ . & . & 1 & . \end{matrix}, \quad S_2^A \rightarrow \begin{matrix} . & -i & . & . \\ i & . & . & . \\ . & . & . & 1 \\ . & . & . & 1 \end{matrix}$$

$$S_1^B \rightarrow \begin{matrix} 1 & . & . & . \\ . & 1 & . & . \\ . & . & . & 1 \\ . & . & 1 & . \end{matrix}, \quad \text{etc.}$$

(see p. 92)

(6) Construct the matrix representing S^{AB} and show that its proper values are $1, 1, 1, -1$.

The Representation of the Vectors of a Composite System.—The appropriate matrix representation of the operators of a composite system can now be deduced as an immediate corollary of the exchange relations established in the last section. To be definite, let us consider a system composed of two particles A and B, and the representation of the two sets of operators

$$N_1^A, N_2^A, N_3^A\,;\; N_1^B, N_2^B, N_3^B.$$

The standard basis for the first three operators is the set of proper vectors $\eta(u_A)$, of N_3^A with proper values $u_A h$; and the standard basis for the second three operators is the set of proper vectors $\eta(u_B)$, of N_3^B with proper values $u_B h$. Similarly, we shall take as the standard basis of both sets of operators the set of simultaneous proper vectors $\eta(u_A\,;\; u_B)$, of N_3^A and N_3^B with simultaneous proper values $u_A h$ and $u_B h$. These three sets of vectors cannot, of course, be pictured in the same space, but we shall establish a relation between the vectors in the spaces **A** and **B** of $\eta(u_A)$ and $\eta(u_B)$, and the vectors in the space **C** of $\eta(u_A, u_B)$.

A system in a definite state can be (partially) represented *either* by a vector α in **A** and a vector β in **B**, *or* by a single vector γ in **C**. Let

$$\alpha = \Sigma\eta(u_A)a(u_A), \quad \beta = \Sigma\eta(u_B)b(u_B),$$

and

$$\gamma = \Sigma\eta(u_A, u_B) \quad c(u_A, u_B).$$

Then $|c(u_A, u_B)|^2$ is the probability that, in the given system, N_3^A and N_3^B have the values $u_A h$ and $u_B h$. But $|a(u_A)|^2$ is the probability that N_3^A has the value $u_A h$, and $|b(u_B)|^2$ is the probability that N_3^B has the value $u_B h$. Now the operators N_3^A and N_3^B commute so that these probabilities are independent. Hence

$$|c(u_A, u_B)|^2 = |a(u_A)b(u_B)|^2.$$

Moreover, by suitably choosing the phases of the components of γ we can ensure that

$$c(u_A, u_B) = a(u_A)b(u_B).$$

This is the required relation between the vectors of **A** and **B** and the vectors of **C**. In view of this relation the space C is described as the " product space " of A and B, and we write

$$\mathsf{C} = \mathsf{A} \times \mathsf{B} = \mathsf{B} \times \mathsf{A}.$$

Now let T^A be the operator representing a variable t^A of the particle A. Then, in the space A, the matrix elements of T^A are given by the equation

$$T^A\eta(u_A') = \Sigma\eta(u_A)T^A(u_A, u_A').$$

To determine the matrix elements of T^A in the product space $\mathsf{A} \times \mathsf{B}$, we note that a determination of the value of t^A will not affect the values of the variables of B. Hence in the space B, t^A is represented by the *identical operator* of this space, I^B, i.e.

$$T^A\eta(u_B') = \eta(u_B') = \Sigma\eta(u_B)I^B(u_B, u_B'), \text{ say,}$$

where $\quad I^B(u_B, u_B') = 0$ if $u_B \neq u_B'$,
or 1 if $u_B = u_B'$.

If we take α to be the vector $T^A\eta(u_A')$ in A, and β to be the vector $T^A\eta(u_B)$ in B, γ will be the vector $T^A\eta(u_A', u_B')$ in $\mathsf{A} \times \mathsf{B}$.
Then

$$T^A\eta(u_A', u_B') = \Sigma\eta(u_A, u_B)T^A(u_A, u_B\,;\; u_A', u_B'),$$

and $\quad T^A(u_A, u_B\,; u_A', u_B') = T^A(u_A, u_A')I^B(u_B, u_B')$
$= 0$ if $u_B \neq u_B'$
or $\quad T^A(u_A, u_A')$ if $u_B = u_B'$.

Similarly, if T^B represents a variable t^B of B, then

$T^B(u_A, u_B\,;\; u_A', u_B') = 0$ if $u_A \neq u_A'$,
or $\quad T^B(u_B, u_B')$ if $u_A = u_A'$.

Moreover, the matrix elements of the sum $T^A + T^B$ in the product space $\mathsf{A} \times \mathsf{B}$ are clearly

$$(T^A + T^B)(u_A, u_B\,;\; u_A', u_B')$$
$$= T^A(u_A, u_A')I^B(u_B, u_B') + T^B(u_B, u_B')I^A(u_A, u_A'),$$

where I^A is the identical operator for the space A.

The representation of $N_1{}^A$, $N_2{}^A$, . . . in terms of the basis $\eta(u_A, u_B)$ follows at once from the general formulæ obtained above, and need not be written out explicitly.

EXAMPLES :—

(1) Show that the proper values of $T^A + T^B$ are the sums of proper values of T^A and of T^B, i.e. are of the form $t_j^A + t_k^B$, and that the proper values of T^AT^B are of the form $t_j^At_k^B$.

(2) Apply the method given above to determine the representation of S_1^A, S_2^A, S_3^A; S_1^B, S_2^B, S_3^B in the product space of the spin variables w^A and w^B.

(3) If the operators $(N_1^A)^2 + (N_2^A)^2 + (N_3^A)^2$ and $(N_1^B)^2 + (N_2^B)^2 + (N_3^B)^2$ have proper values $j_A(j_A + 1)h^2$ and $j_B(j_B + 1)h^2$, $(j_A \geqslant j_B)$, for all the basic vectors $\eta(u_A, u_B)$, show that $N_3^A + N_3^B$ is represented by a diagonal matrix with $(2j_A + 1)(2j_B + 1)$ elements which may be arranged as follows :—

$$\begin{array}{l} j_A + j_B, j_A + j_B - 1, j_A + j_B - 2, \ldots, -j_A - j_B, \\ j_A + j_B - 1, j_A + j_B - 2, \ldots, -j_A - j_B + 1, \\ j_A + j_B - 2, j_A + j_B - 3, \ldots, -j_A - j_B + 2, \\ \cdot \quad \cdot \quad \cdot \quad \cdot \quad \cdot \quad \cdot \quad \cdot \quad \cdot \quad \cdot \quad \cdot \\ j_A - j_B, j_A - j_B - 1, \ldots -j_A + j_B. \end{array}$$

(4) Show that

$$(N_1^A + N_1^B)^2 + (N_2^A + N_2^B)^2 + (N_3^A + N_3^B)^2$$

is represented in the same basis by a diagonal matrix with elements of the form $j(j + 1)h^2$, where

$$\begin{array}{ll} & j = j_A + j_B \; (2j_A + 2j_B + 1 \text{ times}). \\ \text{or} & j_A + j_B - 1 \; (2j_A + 2j_B - 1 \text{ times}), \\ & \cdot \quad \cdot \quad \cdot \quad \cdot \quad \cdot \quad \cdot \quad \cdot \\ \text{or} & j_A - j_B \; (2j_A - 2j_B + 1 \text{ times}). \end{array}$$

The Selection Rules in a Composite System.—The last example shows that even when we possess the maximum information about the separate particles composing a system we are still far from possessing the maximum information about the system as a whole. In order to obtain a complete description of the system as a whole we must include in the fundamental set of commuting operators not only those representing the "partial" momenta, $N_3{}^A$, $K^A = M^A + hI^A$, $N_3{}^B$, K^B, etc., but also the operators

$$N_3 = N^A + N^B + \ldots, \quad M_1{}^2 + M_2{}^2 + M_3{}^2, \quad N_1{}^2 + N_2{}^2,$$

representing the "total" momentum. The proper values of these operators will be denoted by uh, $l(l+1)h^2$ and $j(j+1)h^2$, and it will now be necessary to determine the selection rules for the operators X_1, X_2, X_3 with respect to u, l and j.

If we introduce three operators Σ_1, Σ_2, Σ_3 such that $\Sigma_1{}^2 = I$, $\Sigma_2\Sigma_3 = i\Sigma_1$, etc., and which commute with all the dynamical operators and the spin operators, we may define M and N to be the operators

$$M = M_1\Sigma_1 + M_2\Sigma_2 + M_3\Sigma_3, \quad N = N_1\Sigma_1 + N_2\Sigma_2 + N_3\Sigma_3,$$

and then determine the selection rules by an argument similar to that adopted on page 81 for a single particle. It is obvious that we shall thus obtain the selection rules, $j' = j$ or $j \pm 1$, $l' = l$ or $l \pm 1$, for X_1, X_2 and X_3; and $u' = u$, $u - 1$ or $u + 1$ for X_3, $X' = X_1 + iX_2$ and $X'' = X_1 - iX_2$ respectively. But these results require to be supplemented by two other rules characteristic of a composite system.

In the first place, it is clear that if $j = 0$ and $j' = 0$, then the matrix elements of N', N'' and N_3 will all be zero, and the same will be true of the matrix elements of X', X'' and X_3. Hence we have the additional selection rule

$$\text{if} \quad j = 0, \quad \text{then} \quad j' \neq 0.$$

Secondly, it is necessary to consider the "signature" of a composite state, i.e. the proper value of the "reflection operator,"

$$R = R^A R^B \ldots,$$

where R^A is defined by its action on the wave function

$$\psi(x_1{}^A, x_2{}^A, x_3{}^A, w^A)$$

in the space **A** as

$$R^A\psi(x_1{}^A, x_2{}^A, x_3{}^A, w^A) = \psi(-x_1{}^A, -x_2{}^A, -x_3{}^A, w^A).$$

We have shown (p. 56) that the signature of the particle A is

$$r_A = (-1)^{l_A},$$

$l_A(l_A + 1)h^2$ being the proper value of

$$(M_1^A)^2 + (M_2^A)^2 + (M_3^A)^2.$$

Hence the signature of the composite system is

$$r = r_A r_B \ldots = (-1)^{l_A + l_B + \ldots}$$

(by Ex. 1, p. 100).

Now it is clear that $RX_jR^{-1} = -X_j$, or

$$RX_j + X_jR = 0.$$

Hence the selection rule for R is

$$r' = -r,$$

i.e. $l'_A + l'_B + \ldots$ differs from $l_A + l_B + \ldots$ by an odd number. This is Laporte's rule.

EXAMPLES :—

(1) Show that for a composite system

$$M_1X_1 + M_2X_2 + M_3X_3 \neq 0,$$

and hence show that either X_1, X_2 or X_3 must have matrix elements for which $l' = l$.

(2) Show that the proper values of $\frac{1}{4}(S_1^2 + S_2^2 + S_3^2)$ are of the form $s(s + 1)$ where $s = \frac{1}{2}n$, $\frac{1}{2}n - 1$, . . . 1 or $\frac{1}{2}$, n being the number of particles in the system. (s is called the "spin" quantum number.)

(3) If η is a proper vector of $\frac{1}{2}S_3$, $\frac{1}{4}(S_1^2 + S_2^2 + S_3^2)$, M_3, $M_1^2 + M_2^2 + M_3^2$, N_3, $N_1^2 + N_2^2 + N_3^2$, with proper values, m_s, $s(s + 1)$, $m_l h$, $l(l+l)h^2$, uh, $j(j + 1)h^2$, show that

$$u = m_l + m_s;$$
$$j = l + s, l + s - 1, l + s - 2, \ldots |l - s|;$$
$$u_{max} = l + s, u_{min} = -l - s.$$

Compare the results of Example 2, p. 77, where $n = 1$ and $s = \frac{1}{2}$.

The Symmetry of Systems Containing Similar Particles.—In the preceding study of the matrix representation of the operators of a composite system we restricted ourselves to a consideration of the angular momentum operators. Now such operators serve to characterise a composite system in a partial manner only. A complete description of a composite system would require also the use of the Hamiltonian operators of its sub-systems—or the introduction of the appropriate wave functions—or some equivalent method. But, as soon as we consider the problem of the complete description of a composite system, we encounter a remarkable distinction between systems which do and systems which do not contain " similar " particles.

By " similar " systems in general we mean systems whose Hamiltonian operators are essentially the same: in particular, " similar " particles are particles which have the same mass μ and charge e. Let us consider the special problem of the representation of a system composed of two similar or dissimilar particles by means of Schrödinger's wave functions.

If the particles are dissimilar we may distinguish them as A and B. Then the wave function

$$\psi(x_1, x_2, x_3\,;\ y_1, y_2, y_3) = \psi(x\,;\ y)$$

determines the probability that A is in the volume element $dx_1\, dx_2\, dx_3$ at x_1, x_2, x_3 and that B is in the volume element $dy_1\, dy_2\, dy_3$ at y_1, y_2, y_3. This probability is given by

$$|\psi(x, y)|^2\, dx_1\, dx_2\, dx_3\, dy_1\, dy_2\, dy_3,$$

and it is quite distinct from the probability,

$$|\psi(y, x)|^2\, dy_1\, dy_2\, dy_3\, dx_1\, dx_2\, dx_3,$$

that A is in the volume element at y_1, y_2, y_3, and that B is in the volume element at x_1, x_2, x_3. The interchange of two dissimilar particles will produce an observable change in the state of the composite system. (If we are using spin operators we must include the spin variables in the argument of the wave function.)

If, however, the particles are similar we can distinguish them only by the positions which they occupy.

Hence the wave function $\psi(x\,;\,y)$ determines no more than the probability that *one of the two particles* is at x_1, x_2, x_3, and that *the other of the two particles* is at y_1, y_2, y_3. As before, this probability is given by

$$|\psi(x, y)|^2\, dx_1\, dx_2\, dx_3\, dy_1\, dy_2\, dy_3,$$

which is equally the probability that one particle is at y_1, y_2, y_3 and the other at x_1, x_2, x_3. Hence

$$|\psi(x, y)|^2 = |\psi(y, x)|^2,$$

and, moreover, the wave functions $\psi(x, y)$ and $\psi(y, x)$ must refer to the *same* state (there is no observational criterion for the interchange of similar particles), i.e.

$$\psi(y, x) = e^{ia}\psi(x, y),$$

and $$\psi(x, y) = e^{-ia}\psi(y, x).$$

The phase " a " may possibly be a function of x and y, $a(x, y)$. But, if so, it must satisfy the condition

$$a(x, y) = -\,a(y, x),$$

and it must be invariant for any translation or rotation of the co-ordinate system. The only such invariant function of x and y is the distance

$$(y_1 - x_1)^2 + (y_2 - x_2)^2 + (y_3 - x_3)^2,$$

which does not satisfy the first condition. Hence the phase " a " must be independent of x and y. Therefore

$$\psi(x, y) = e^{ia}\psi(y, x),$$

i.e. $$e^{ia} = e^{-ia},$$

or $$e^{ia} = +\,1 \text{ or } -\,1.$$

The wave function for a system composed of two similar particles is therefore either *symmetrical* or *anti-symmetrical* in the co-ordinates of the two particles, i.e.

either $$\psi(x, y) = \psi(y, x),$$

or $$\psi(x, y) = -\,\psi(y, x).$$

To determine which of these mathematical possibilities is actually realised we must appeal to the empirical facts of atomic and molecular structure, which can be correlated with the general principles of quantum theory, only by assuming that *anti-symmetry* is characteristic of pairs of electrons or of pairs of protons. Hence for electrons or protons we must use anti-symmetrical wave functions.

In the general case of a composite system containing m electrons and n protons, a similar argument allows us to conclude that the wave function must be anti-symmetrical in each pair of electron co-ordinates, and anti-symmetrical in each pair of proton co-ordinates.

EXAMPLES :—

(1) If P is the (permutation) operator which interchanges the co-ordinates of a certain pair of similar particles in the wave function of a composite system, and if W is the Hamiltonian operator of the system, show that P and W commute.

(2) The m, n-element in a determinant Δ is the function $\psi_m(x_n)$. Prove that Δ is anti-symmetrical in each pair of arguments, x_j, x_k. Construct the analogous function of the ψ_m's which is symmetrical in each pair of arguments.

(3) Let W_x, W_y be differential operators which act on the variables x, y respectively. If P is the permutation operator defined by

$$Pf(x, y) = f(y, x),$$

prove that $PW_x = W_yP$; $PW_y = W_xP$,

and $P(W_x + W_y) = (W_x + W_y)P$.

(4) If ψ is the wave function for a system composed of similar sub-systems, each containing j electrons and k protons, show that ψ will be symmetrical or anti-symmetrical in the co-ordinates of each pair of sub-systems accordingly as $j + k$ is even or odd, i.e. according as the net charge on each sub-system is an even or odd multiple of the electronic charge.

The Interaction Energy and Perturbation Theory.— In classical dynamics the total energy of a system composed of sub-systems A, B, . . . N is the sum of the proper (or self-) energies of the individual systems, w^A, w^B, . . . , and the mutual (or interaction) energies of the systems taken in pairs, w^{AB}, w^{AC}, w^{BC}, . . . , so that

$$w = w^A + w^B + \ldots + w^{AB} + w^{AC} + \ldots$$

Similarly, we assume that in quantum dynamics the Hamiltonian operator W of a composite system is the sum of a number of operators, W^A, W^B, . . . W^{AB}, . . . of which W^A is the Hamiltonian operator of sub-system A, . . . etc., and W^{AB} the operator representing the interaction energy of sub-systems A and B, . . . etc. The sum of the operators W^{AB}, . . . , is called the interaction operator U. Clearly W^A commutes with W^B, etc., but does not in general commute with W^{AB} or U.

To overcome the difficulties which arise from the existence of an interaction operator U which does not commute with W^A, W^B, . . . , the following device is frequently employed in describing composite systems :—

We consider the composite system whose interaction operator is λU, where λ is a real numerical constant, and we show that all the numbers—proper values of operators, values of transition probabilities, average values of variables—which characterise a definite state of the composite system are expressible as power series in λ with coefficients completely determined by their leading terms which, of course, correspond to the value $\lambda = 0$. That is to say, the properties of the actual system for which $\lambda = 1$ can be mathematically deduced from the properties of the ideal system, for which $\lambda = 0$. The details of the mathematical technique, " the perturbation theory," are irrelevant to our general line of exposition, and will not be discussed here. The conclusion which we borrow from the mathematical analysis is that the actual system ($\lambda = 1$) is completely and adequately described (in an implicit manner) by the ideal system ($\lambda = 0$), which may be considered to evolve into the actual system as λ increases from 0 to 1.

Consider, for example, two similar one-dimensional harmonic oscillators which are elastically coupled. Let the oscillator A have kinetic energy $\frac{1}{2}\mu\dot{x}_A{}^2 = p_A{}^2/2\mu$ and potential energy $\frac{1}{2}\mu\omega_0{}^2x_A{}^2$ when its displacement from its position of equilibrium is x_A. Then

$$2\mu W^A = (P^A)^2 + \mu^2\omega_0{}^2(X^A)^2,$$

and, similarly,

$$2\mu W^B = (P^B)^2 + \mu^2\omega_0{}^2(X^B)^2.$$

The proper values of W^A, W^B are both of the form $(n + \frac{1}{2})h\omega_0$, where n is a positive integer or zero (see p. 61). Let the mutual energy of A and B be

$$\tfrac{1}{2}\mu c(x_A + x_B)^2.$$

Then $$U = W^{AB} = \tfrac{1}{2}\mu\omega_0{}^2c(X^A + X^B)^2.$$

We shall now obtain the proper values and the wave functions of the operator $W^A + W^B + \lambda U$ as functions of the parameter. On introducing the "principal" co-ordinates x_1 and x_2 defined by the equations

$$\sqrt{2}\,.\,x_1 = x_A + x_B, \quad \sqrt{2}\,.\,x_2 = x_A - x_B,$$

the total kinetic energy operator, $[(P^A)^2 + (P^B)^2]/2\mu$, becomes

$$\tfrac{1}{2}\mu[(X^A)^2 + (X^B)^2] = \tfrac{1}{2}\mu[(X_1)^2 + (X_2)^2] = (P_1{}^2 + P_2{}^2)/2\mu,$$

and the total potential energy operator,

$$\tfrac{1}{2}\mu\omega_0{}^2[(X^A)^2 + (X^B)^2] + \tfrac{1}{2}\lambda\mu\omega_0{}^2c(X^A + X^B)^2,$$

becomes $$\tfrac{1}{2}\mu\omega_0{}^2(X_1{}^2 + X_2{}^2) + \lambda\mu\omega_0{}^2cX_1{}^2.$$

Hence the Hamiltonian operator is now

$$[P_1{}^2/2\mu + \tfrac{1}{2}\mu\omega^2(\lambda)X_1{}^2] + [P_2{}^2/2\mu + \tfrac{1}{2}\mu\omega_0{}^2X_2{}^2],$$

where $$\omega^2(\lambda) = \omega_0{}^2(1 + 2\lambda c).$$

It is easily verified that the two parts, W_1 and W_2, into which we have divided the Hamiltonian operator, will commute with one another. Hence the composite system is adequately described by the proper values of

W_1 and W_2, say $(m + \frac{1}{2})h\omega(\lambda)$ and $(n + \frac{1}{2})h\omega_0$. Also, if $\psi[x_1, (m + \frac{1}{2})h\omega_0]$ and $\psi[x_2, (n + \frac{1}{2})h\omega_0]$ are the wave functions of W_1 and W_2 when $\lambda = 0$, the wave functions in the general case are

$$\psi[x_1(m + \tfrac{1}{2})h\omega(\lambda)] \quad \text{and} \quad \psi[x_2, (n + \tfrac{1}{2})h\omega_0],$$

i.e. they are completely determined by the wave functions for the ideal case $\lambda = 0$. Hence the state of the actual system $(\lambda = 1)$ is completely determined by the ideal state $(\lambda = 0)$.

It is this device which is employed in the classification of atomic structures. The actual atom is replaced, for purposes of classification, by an ideal atom in which there is no interaction between the different electrons, although the interaction of the electrons with the nucleus is retained unaltered. Each electron is therefore treated as if it moved under the attraction of the nucleus alone, and the proper values of its energy operator are therefore the energy levels of hydrogen-like atoms. Hence the ideal atom can be completely described by the proper values of the angular momentum and energy operators of the electrons of which it is composed, and these numbers also serve to describe the states of the actual atom.

EXAMPLES :—

(1) Let $\eta_1, \eta_2, \ldots$ be simultaneous proper vectors of the self energies, $W^A, W^B \ldots$ of a composite system. Prove that the probability that the system should pass from the state η_j at time $t = 0$ to the state η_k at time $t = \tau$ is $|(\eta_k, F\eta_j)|^2$, where $F = \exp(-iU\tau/h)$, U being the interaction operator.

(2) If $\psi(x, \lambda)$ is the wave function for the system with Hamiltonian operator $W = H + \lambda U$ and if

$$\psi(x, \lambda) = \psi_0(x) + \lambda\psi_1(x) + \ldots + \lambda_n\psi_n(x) + \ldots,$$

prove that

$$H\psi_0(x) = 0, \quad H\psi_1(x) = -U\psi_0(x), \ldots, \\ H\psi_{n+1}(x) = -U\psi_n(x), \ldots .$$

(3) Let $\psi_1(x)$, $\psi_2(x)$, . . . be a complete set of wave functions representing a set of proper states of a certain system with Hamiltonian operator W. Show that the wave function

$$\psi = \sum_n \psi_n(x)c_n \,.\, \exp(-iwt/h)$$

will represent a stationary state if c_1, c_2, . . . are the components of a proper vector γ, of W with proper value w.

(4) Determine the most general form of ψ and the form of the probability that the states represented by ψ and ψ_n should be the same.

$$[|\Sigma c_{n,k} \exp(-iw_k t/h)|^2.]$$

The Exclusion Principle.—We must now consider in more detail the representation of the stationary states of an "ideal" composite system. The Hamiltonian operator of the complete system is the sum of the Hamiltonian operators of the sub-systems which compose it, and these operators obviously commute with one another. Hence a stationary state will be (at least partially) described as a proper state of the operators W^A, W^B, . . . W_N. This description in terms of the partial energies, w_A, w_B, . . ., must be completed by employing additional groups of variables characteristic of the individual sub-systems, e.g. the angular momentum variables. A stationary state of the composite system will then be completely specified as a proper state of N groups of operators, say

$$W^A, Q_1^A, Q_2^A, \ldots;\ W^B, Q_1^B, Q_2^B, \ldots;\ \ldots$$

Let the individual sub-systems, A, B, . . ., be represented in their own proper spaces **A**, **B**, . . . by the wave functions $\psi(x_A, q_j)$, $\psi(x_B, q_k)$, . . . where x_A summarises the positional and spin co-ordinates of A, etc., and q_j summarises the proper values of W^A, Q_1^A, Q_2^A, . . . for the wave function $\psi(x_A, q_j)$, etc. If the sub-systems are all dissimilar the composite system is represented in the product space $\mathsf{A} \times \mathsf{B} \times \ldots$ by wave functions of the form

$$\psi(x_A, q_j), \psi(x_B, q_k) \ldots \text{ (a product of N factors),}$$

there being no restrictions on the factors employed. But if the sub-systems are all similar the wave function representing the composite system must be symmetrical or anti-symmetrical in each pair of co-ordinates x_A, x_B, ... If the sub-systems are electrons the wave function must be anti-symmetrical and will therefore have the form of a determinant in which $\psi(x_M, q_n)$ is the element in the Mth row and n^{th} column.

This determinant will clearly vanish identically if any two sets of proper values, say q_j and q_k, are the same, for, in this case, the jth and k^{th} columns of the determinant would be identical. Now the sets of proper values, $q_j, q_k, \ldots$ serve to specify the state of the ideal composite system in terms of states of the sub-systems which can be regarded as "generating" the composite system. Hence the only states of the ideal composite system which are actually realised are those generated by N *different* states of the N sub-systems. This is Pauli's exclusion principle. It shows that the anti-symmetry characteristic of systems composed of similar systems excludes from consideration all states of the composite system in which any pair of the generating sub-systems are in the same state.

EXAMPLES :—

(1) Let P denote any permutation of the N groups of co-ordinates x_A, x_B, ..., x_N, and let Px_A, Px_B, ..., Px_N denote the groups by which they are replaced. Show that the operator P defined by the equation

$$P\psi(x_A, x_B, \ldots) = \psi(Px_A, Px_B, \ldots)$$

is a linear, unitary operator which commutes with the Hamiltonian operator of a system of N similar particles, provided that the interaction is symmetrical in each pair of particles.

(2) Let $P_1 = I, P_2, P_3, \ldots, P_{N!}$ denote the N! permutations of the N groups of co-ordinates $x_A, x_B, \ldots, x_N$. Show that the operator

$$\Pi_A = (1/N!) \sum_n P_n P_A P_n^{-1} \quad (n = 1, 2, 3, \ldots N!)$$

commutes with all the P_j's, and that the operators $\Pi_A, \Pi_B, \ldots$ all commute with one another.

(3) Show that the probability of a transition from any proper state of $\Pi_A, \Pi_B, \ldots$ to any other proper state is rigorously zero, i.e. the proper states of these operators are entirely independent.

(4) Show that in a symmetrical composite system the proper values of the Π's are all $+1$, and that in an anti-symmetrical system the proper values are ± 1.

The Electron Configuration of Ideal Atoms.—To apply Pauli's exclusion principle to an atom containing N similar electrons it is only necessary to decide upon a standard set of operators for the description of the electronic states which "generate" the state of the whole atom. A suitable set of operators (for electron A) is

$$W^A,\ (N_1^A)^2 + (N_2^A)^2 + (N_3^A)^2,\quad (M_1^A)^2 + (M_2^A)^2 + (M_3^A)^2,\ N_3^A.$$

The proper values of these operators are of the form

$$-CZ^2/n^2,\ j(j+1)h^2,\ l(l+l)h^2,\ uh,$$ (see p. 88),

where $l = 0, 1, 2, \ldots, (n-1)$,

$j = (l + \frac{1}{2})$ or $(l - \frac{1}{2})$,

and $u = -j, -j+1, \ldots, j-1, j$.

The electron configuration of an atom is then specified by N *different* sets of quantum numbers, each set of the form (n, l, j, u).

If we consider an "ideal" atom, ignoring the interaction energy, which may well be far from negligible, the total energy of the atom will be

$$-CZ^2.\{n_A^{-2} + n_B^{-2} + \ldots + n_N^{-2}\}.$$

An atom is said to be in its " normal " state when the absolute value of its total energy is as large as possible; and the determination of the normal state of an ideal atom provides a simple illustration of the exclusion principle.

It is clearly unnecessary that all the " principal " quantum numbers, n_A, n_B, . . . , should be different, and in the normal state we must make as many as possible equal to 1, then as many as possible equal to 2, and so on.

The electrons for which $n = 1$ are said to form the " K-shell." For these electrons

$$n = 1, \quad l = 0, \quad j = \tfrac{1}{2}, \quad u = \pm \tfrac{1}{2}.$$

Hence the maximum number of electrons in the K-shell is 2.

The electrons for which $n = 2$ form the " L-shell." For these electrons

$$n = 2, \begin{cases} l = 0, j = \tfrac{1}{2}, \ u = \pm \tfrac{1}{2} \\ \text{or} \\ l = 1, \begin{cases} j = \tfrac{1}{2}, u = \pm \tfrac{1}{2}, \\ \text{or} \\ j = \tfrac{3}{2}, u = \pm \tfrac{1}{2}, \pm \tfrac{3}{2}. \end{cases} \end{cases}$$

Hence the maximum number of electrons in the L-shell is 8.

Similarly, we can show that in any " sub-group," in which each electron has the same principal quantum number n and the same serial quantum number l, the maximum number of electrons is 2 if $l = 0$ and

$$(2|l + \tfrac{1}{2}| + 1) + (2|l - \tfrac{1}{2}| + 1) = 2(2l + 1) \text{ if } l \neq 0,$$

i.e. in both cases the maximum number is $2(2l + 1)$. This is *Stoner's rule*. It follows that in any complete shell or group in which each electron has the same principal quantum number n, the maximum number of electrons is

$$\Sigma 2(2l + 1), [l = 0, 1, 2, \ldots, (n - 1)], = 2n^2.$$

The further developments of the theory of atomic structure are beyond the scope of this book. The main difficulties rise, as we should expect, in making the transition from ideal atoms to actual atoms, but they are difficulties of method rather than of principle.

EXAMPLES :—

(1) Show that the "normalised" wave function for a system of N electrons is $(N!)^{-\frac{1}{2}} \times$ the determinant of the wave functions $\psi(x_A, q_j)$ of the electronic states.

(2) Show that for a complete shell the proper values of $S_3 \equiv S_1^A + S_2^A + \ldots$, M_3, N_3, $M_1^2 + M_2^2 + M_3^2$, $N_1^2 + N_2^2 + N_3^2$ are all zero.

[There is only one possible configuration for a complete shell, therefore only one wave function, which is therefore a proper vector of all the above operators, etc.]

Quantum Statistics.—The outline of the general principles of the quantum theory sketched in this monograph may conveniently be brought to a close by a brief account of the application of these principles to statistical mechanics. Here the subject of study is a collection of similar material sub-systems (such as a gas) which do not interact with one another directly, but only through the medium of radiation. Hence the complete system must consist of the material sub-systems *plus* the radiation, and its Hamiltonian operator will have the form

$$W = W^A + H^A + W^B + H^B + \ldots + R,$$

where R is the Hamiltonian operator of the radiation and H^A the operator representing the energy of interaction of the sub-system A with the radiation. Such a system is most conveniently described in terms of the proper states common to W^A, W^B, . . . , R, and the main problem is to calculate the probability that a sub-system will pass from one of its proper states "*a*" to another proper state "*b*" This transition probability

(p_{ab}) depends upon the matrix elements U_{JK} of the interaction operator

$$U = H^{A} + H^{B} + \ldots,$$

and we shall carry our calculations sufficiently far to compare the two probabilities p_{ab} and p_{ba} of two reciprocal transitions ($a \rightarrow b$, $b \rightarrow a$), which, we shall find, are generally different in value.

We shall consider the case of a symmetrical system, and we shall ignore the operator R and its proper vectors —the modifications necessary when R is retained can be supplied without difficulty. Let $Q \equiv (q_1, q_2, \ldots, q_a, \ldots)$ be the complete array of all possible sets of quantum numbers for any one of the sub-systems. (Since the sub-systems are similar, this array is the same for every sub-system.) Let J be a collection of N sets of quantum numbers (with repetitions if desired !), say $j_1, j_2, \ldots, j_N$, selected from the complete array Q, and let $J_1 = I_1, J_2, \ldots J_m, \ldots$ be the group of all possible permutations of these N sets of quantum numbers. Then, if $\psi(x_A, j_1)$, $\psi(x_B, j_2)$, . . . , are wave functions representing proper states of the individual sub-systems, A, B, . . . , a wave function of the complete system will be represented by a sum of the form

$$\Psi_J = a_J \,.\, \Sigma_m \, \psi(x_A, J_m j_1)\psi(x_B, J_m j_2) \ldots,$$

extended over all permutations of the collection J. (Here, as on p. 110, $J_m j_n$ represents the set of quantum numbers by which the permutation J_m replaces j_n.)

This wave function can also be specified by the set of numbers $n_1, n_2, \ldots, n_a, \ldots$, in which n_a denotes the number of j's in J which are equal to q_a; and we may write

$$\Psi_J = \Psi(n_1, n_2, \ldots).$$

The numerical coefficient a_J is to be chosen so as to normalise Ψ_J. Now, if $\int \ldots dv$ denotes the operation of integrating and summing over the domain of all the positional and spin co-ordinates of A, B, . . . , then

$$\begin{aligned}\int \Psi_J{}^*\Psi_J dv &= a_J{}^* a_J \,.\, \sum_{m,n} \int \psi^*(x_A, J_m j_1)\psi(x_A, J_n j_1) \ldots dv \\ &= |a_J|^2 \,.\, \sum_m \int \psi^*(x_A, J_m j_1)\psi(x_A, J_m j_1) \ldots dv \\ &= |a_J|^2 \times \text{the number of permutations of the collection J} \\ &= |a_J|^2 \times N!/n_1!\,n_2!\ldots\end{aligned}$$

Hence we may take a_J to be $(n_1!\,n_2!\ldots/N!)^{\frac{1}{2}}$.

In calculating the matrix elements of U we shall require to know the number of permutations of the collection J which replace a certain number by one of the n_a numbers equal to q_a. This is clearly

$$(N-1)!/n_1!\,n_2!\ldots(n_a-1)!\ldots$$

or $n_a/N \times$ the total number of these permutations.

We also note that the matrix elements of any operator H^A in the space **A**, which are given by the integrals

$$\int \psi^*(x_A, q_a) H^A \psi(x_A, q_b) dv_A,$$

are independent of the affix " A " which distinguishes the individual systems, and may therefore be denoted by the symbols H_{ab}.

We can now calculate the matrix elements of H^A which correspond to a transition from the state of the complete system represented by Ψ_J to the state represented by a similar type of wave function Ψ_K. We have that

$$H_{JK}{}^A = \int \Psi_J{}^* H^A \Psi_K dv$$

$$= a_J{}^* a_K \,.\, \sum_{m,n} \int \psi^*(x_A, J_m j_1) H^A \psi(x_A, K_n k_1) \psi^*(x_B, J_m j_2) \psi(x_B, K_n k_2) \ldots dv.$$

Now the integrals

$$\int \psi^*(x_B, J_m j_2)\psi(x_B, K_n k_2) dv_B, \ldots,$$

all vanish unless

$$J_m j_p = K_n k_p \quad \text{for} \quad p = 2, 3, \ldots N,$$

in which case their product is unity. Hence all the matrix elements of H^A will vanish unless the two collections of sets of quantum numbers

$$J_m(j_2, j_3, \ldots j_N) \text{ and } K_n(k_2, k_3, \ldots k_N),$$

are identical for some m and n, i.e. the collections $(j_2, j_3, \ldots j_N)$ and $(k_2, k_3, \ldots k_N)$ are identical. Hence we may write the collections J and K in the form

$$J \equiv (q_a, q_l, q_m, \ldots) \text{ and } K \equiv (q_b, q_l, q_m, \ldots).$$

If we write

$$\Psi_J = \Psi(n_1, n_2, \ldots) \text{ and } \Psi_K = \Psi(n_1', n_2', \ldots),$$

the selection rules for H^A are clearly

$$n_a' = n_a - 1, \; n_b' = n_b + 1,$$

but, otherwise,

$$n_p' = n_p.$$

Hence each matrix element of H^A refers to a " simple " process in which a single sub-system leaves the state specified by the set of quantum numbers q_a and enters the state specified by the set q_b.

For the non-vanishing matrix elements the only surviving terms in the expression for $H_{JK}{}^A$ are

$$a_J{}^* a_K \,.\, \Sigma \int \psi^*(x_A, q_a) H^A \psi(x_A, q_b) dv_A,$$

the sum being taken over all permutations which leave q_a (or q_b) unaltered.

Hence

$$H^A_{JJ} = \Sigma[n_a/N] \,.\, H_{aa},$$

and

$$H^A_{JK} = [(n_b + 1)/n_a]^{\frac{1}{2}} \,.\, [n_a/N] \,.\, H_{ab}, \quad (J \neq K).$$

But

$$U_{JK} = H^A_{JK} + H^B_{JK} + \ldots = N \,.\, H^A_{JK}.$$

Therefore,

$$U_{JJ} = \Sigma n_a \,.\, H_{aa}$$

and

$$U_{JK} = [n_a(n_b + 1)]^{\frac{1}{2}} \,.\, H_{ab}, \quad (J \neq K).$$

Hence the probability of a transition in which a sub-system passes from the state specified by q_a to the state specified by q_b in (a short) time t is (see p. 65),

$$p_{ab} = |U_{JK}|^2 \,.\, t^2/h^2 = n_a(n_b + 1)|H_{ab}|^2 t^2/h^2 \,;$$

and similarly, the transition probability for the converse process is

$$p_{ba} = n_b(n_a + 1)/H_{ab}/^2t^2/h^2.$$

EXAMPLES :—

(1) Let N_a, M_a, P_a, . . . be operators defined by the relations

$$N_a\Psi_J = n_a\Psi_J, \ldots,$$
$$M_a(N_a + 1) = (N_a)M_a, \ldots,$$
$$P_a(N_a) = (N_a + 1)P_a, \ldots,$$
$$N_a, M_a, P_a \text{ commute with } N_b, M_b, P_b, \ldots$$

Show that $U = \sum_{a,b} N_a^{\frac{1}{2}}M_a \,.\, H_{ab} \,.\, P_bN_b^{\frac{1}{2}}$.

(These are the operators employed in "superquantisation," M_a, P_a usually being written as

$$e^{iW_a},\ e^{-iW_a}.)$$

(2) It may be assumed that the effect of including the radiation in the preceding argument is to prove (1) that the transition of a material sub-system from a state q_a to a state q_b is accompanied by the transition of a radiative sub-system from a state $q_{a'}$ to a state $q_{b'}$, the transition probability, $p_{aa',\,bb'}$, being proportioned to

$$n_an_{a'}(n_b + 1)\,(n_{b'} + 1)\,;$$

and (2) that there is conservation of energy, i.e.

$$E_a + E_a' = E_b + E_b'.$$

If, in a state of equilibrium,

$$p_{aa',\,bb'} = p_{bb',\,aa'},$$

and $\qquad n_a = f(E_a), \quad n_a' = f'(E_a')$, etc.,

show that

$$n_a = 1/(Ae^{\mu W_a} - 1), \quad n_a' = 1/(Ae^{\mu W_a} - 1).$$

Hence infer the form of Planck's formula for complete radiation.

(3) If Ψ_J is an anti-symmetrical function, show that it can be written in the form $\Psi(n_1, n_2, \ldots)$ where each $n_a = 1$ or 0, and show that the selection rules for U are

$$n_a = 1 \rightarrow n_a' = 0,$$
$$n_b = 0 \rightarrow n_b' = 1,$$

and otherwise $n_p' = n_p$.

GENERAL REFERENCES

"The Principles of Quantum Mechanics," P. A. M. Dirac. (Oxford University Press, 1930.)

"The Theory of Groups and Quantum Mechanics," H. Weyl. (Methuen & Co., 1931.)

"Mathematische Grundlagen der Quantenmechanik," J. v. Neumann. (J. Springer, Berlin, 1932.)

"The Physical Principles of the Quantum Theory," W. Heisenberg. (Cambridge University Press, 1930.)

INDEX